Who can legitimately be considered a technologist, designer or engineer?

Anne Galloway
p.22

What changes are we unknowingly ushering in as we scroll, tap, browse and swipe for hours each day?

Corinna Gardner
p.28

When my robot learns to love, will I be able to love it back?

Mariana Pestana
p.34

What a relief not to have to sit across from her any more.

The mayor donned the red and yellow Lycra outfit. This gesture was meant to set an example of what responsible citizenship might entail.

If the crowdfunding model of participation could be harnessed, then the cities of the future might look quite different.

Kieran Long
p.60

When your world is destroyed in front of your eyes, what do you do?
How do you react?
How do you survive?
Can you remain hopeful?
Can you persevere?
Can you rebuild?

Alex Kalman
p.64

What does a drone have to do with the internet?

Teju Cole
p.72

How the next half of the world connects to the internet will look very different from how the first half has connected.

Designing the perfection of violence?

Although it is not quite right to compare the internet to a forest, the analogy can help to understand the nature of complex systems.

Natalie D. Kane
p.100

How to reinvent the nuclear accident is perhaps the greatest design challenge of all.

Susan Schuppli
p.104

Who would not rather combat terrorism with a tree than a drone?

Jennifer Kabat
p.110

But what is there to be done in space, except sit tight (or float about) and wait for a replacement to make the long, expensive journey from Earth?

Zara Arshad
p.120

Is your burger really beef?

Alexandra Daisy Ginsberg
p.136

Many people who today are regarded as dying may in the future be in a salvageable state. If only they could end up in that future.

Anders Sandberg
p.142

What is the point of a computer that can play videogames?

Rory Hyde
p.148

The future emerging through design is a sort of religious future, although it appears in the guise of technology.

Arjun Appadurai
p.154

THE FUTURE STARTS HERE

Edited by Rory Hyde and Mariana Pestana with Kieran Long
Assistant Editor Zara Arshad

V&A Publishing

Published to accompany the exhibition *The Future Starts Here*
at the Victoria and Albert Museum, London
(12 May–4 November 2018)

Supported by

VOLKSWAGEN
GROUP

First published by V&A Publishing, 2018
Victoria and Albert Museum
South Kensington
London SW7 2RL
www.vandapublishing.com

Distributed in North America by Abrams, an imprint of ABRAMS

ISBN 9781 85177 907 9
10 9 8 7 6 5 4 3 2 1
2021 2020 2019 2018

A catalogue record for this book is available from the British
Library.

Designer: JULIA (julia.studio)
Copy-editor: Linda Schofield
Printed in Italy

V&A Publishing
Supporting the world's leading
museum of art and design,
the Victoria and Albert
Museum, London

CONTENTS

FOREWORD

Caring about tomorrow while paying attention to today is as essential for individuals as it is for companies, especially when it comes to balancing technology and life. *The Future Starts Here* examines such a balance and encourages visitors to explore possible future realities.

As part of the process of becoming a leading global provider of sustainable mobility, the Volkswagen Group is actively involved in shaping the future by introducing pioneering technologies and fulfilling its commitment as a strong partner for customers and employees worldwide. This progress is driven by our sustained appreciation for creativity, which powers engineers and artists alike. Through our partnerships, including our collaboration with the V&A, we encourage people to develop their imagination and inventiveness. The courage to generate new ideas is inspired by experiencing a variety of creative work – that is why the Volkswagen Group focuses so much of its social responsibility on the support of cultural institutions and artists.

We would like to thank the V&A and its dedicated staff for their cooperation and the inspiring exchange of views. They paved the way for this exceptional exhibition, which brings tomorrow to the here and now.

Benita von Maltzahn
Volkswagen Group

FOREWORD

The Future Starts Here explores the power of design in shaping the world of tomorrow. It interrogates groundbreaking emerging technologies, the ways they will affect our lives in the near future, and the collective choices we have to influence their progress. The exhibition delves into the fast-accelerating future of DNA analysis, artificial intelligence, synthetic biology and space exploration now emerging from studios and laboratories across the globe.

From its earliest foundation, the V&A has brought together cutting-edge objects at the forefront of art, design, science and technology. Now 160 years on, and in the midst of the digital revolution, we are embracing the new frontiers arising from the cross-section of design and science. This exhibition dives into the astonishing world of future-facing design to project our possible futures, taking live experiments and real objects from the studio and lab into the museum. This is the first time that many of these works have been shown in this environment: a defining moment for the museum, as mediator and collaborator in these revolutionary conversations.

The studies and stories brought together in this volume expand these conversations with the help of a truly impressive range of authors, designers and thinkers. Each of these 'object essays', as the editors call them, reveals the extreme depth and breadth of connections contained even within the simplest objects of today.

It gives me great pleasure to thank our sponsor, without whom this exhibition would not have been possible. The V&A is delighted to work with Volkswagen Group as the lead exhibition supporter, a fitting partner given its own ethos of forward-looking design and research.

Tristram Hunt
Victoria and Albert Museum

The Great Exhibition of 1851 brought art,
design, technology and nature together in
a flat hierarchy.

INTRODUCTION

The V&A was born out of the idea to host a collective event to make sense of the future. The Museum's origins trace back to the first Great Exhibition staged in the Crystal Palace, London, in 1851. This exhibition brought together the greatest achievements of the world, combining fine arts with new technologies, machinery and even raw materials. Presented in a flat hierarchy, examples of art, design and technology were displayed alongside one another, providing an overarching picture of the scientific and cultural innovations that would usher in the modern world. Many of these objects formed the basis of the collection of the South Kensington Museum as a means of preserving this one-off event in an enduring public institution.

Some decades later, at the end of the nineteenth century, this all-encompassing structure had become unwieldy, and two new museums were created to provide some distinction. Reflecting the logic of the age, obsessed as it was with categorization, the collection was divided into distinct specialisms: the machines and technology were transferred to the Science Museum, while the art and design were kept in the greatly expanded South Kensington Museum, now named the V&A. Thus this brief period of a unified approach to creativity came to a close.

Albeit on a more modest scale, *The Future Starts Here* seeks to reunite the fields of art, design, science and technology, after more than a century of separation. By cutting across these disciplines and placing them in dialogue, we once again reflect the contemporary reality of creative production, and are able to draw a new picture of an emerging future. If the Great Exhibition of 1851 sought to make sense of a world in the midst of the Industrial Revolution, *The Future Starts Here* assembles technologies and products in order to make sense of the nascent digital revolution. These are real objects that are either newly released or in development, produced by designers, universities, corporations, governments and collectives. At present these objects exist within the limited confines of the lab or studio, but what would happen if they were to be widely adopted? How might they change the way we live, the way we work, the way we love?

This book is organized into four sections that evoke scales of technological impact: self, public, planet and afterlife. Each section is introduced by a series of images of projects that evoke and engage with these scales. For the essays that follow, we assigned each author an object from the exhibition, and invited them to use it as a starting point. The imaginative leap of the authors is to envisage what futures might be contained within these objects, and how they might transform our everyday lives. They also bring different perspectives: not just technological, but emotional, personal, physical and metaphysical, revealing the unique depth and complexity that these objects contain. None of them are as straightforward as they appear, but instead sit at the intersection of a vast network of vectors: manufacturing, globalization, sociability, economics and design.

Looking at this parade of objects, it is tempting to extrapolate forward, to trust their power and inevitability and use them to see the future. But the future is less predictable than that. These objects are merely claims, attempts by their creators to bring a particular future into being. Their undeniable physical reality – their *objectness* – may give the impression that this future has already arrived, complete and self-contained. But while these objects may be predisposed towards certain outcomes, neither are they straitjackets, as they

8

cannot account for their users. For us. They can be used or misused, supported or ignored, promoted or hacked, resulting in an unpredictable combination of the good and the bad, utopia and dystopia, triumph and failure. As Paul Virilio reminds us, the invention of the ship was also the invention of the shipwreck.[1] New things contain new potentials and possibilities, often unknown and unanticipated even by their creators. Thus the objects presented here are nothing more than best guesses, experiments hoisted into the world in the hope they will somehow change it. We can sit back and let that happen, or we can take an active role in steering where these things take us. The future then becomes a collective project for all of us, to work together to pull along the ideas we want, and to push against those unwillingly foisted upon us. Despite the inherent unpredictability of this point in time – a fork in the vectors of past and future – this lack of resolution should be grasped as an opportunity rather than a cause of anxiety.

In the course of putting this book and exhibition together we have met many stunningly intelligent people who are invested in the idea of the future, by conceiving it, researching it, designing it, building it or even selling it. They can see a bit further over the horizon than the rest of us, and yet despite this view of the various catastrophes that could lie ahead of us, they are not dispirited or disheartened, but remain enthusiastically committed. They all seem to believe in the future, in wanting to make it better, and in their ability to shape it. When it would be much easier to ignore, to look away, these people are more engaged. They are engaged because despite our future appearing determined, they know there is still much to play for. They know that the future starts here.

Rory Hyde and Mariana Pestana

SELF

Recent developments in biology and robotics are transforming our perceptions of what makes us human. We can edit the strands of DNA that make up our genetic code, and have invented implants and devices that enhance our 'natural' capabilities. At the same time, we are writing codes for bringing machines to life through artificial intelligence. How are these new fields of design challenging our self-perception?

Body Code
Drew Berry
2017

This scientifically accurate illustration by biological animator Drew Berry depicts the world inside a human cell. It shows the 'mechanical clockwork' of a chromosome, a component of the genetic code that makes up life. Allowing us to visualize DNA, Berry's work unlocks a new visual language and a future site of design.

Self

Solar Shirt
Pauline van Dongen
2015

Flexible solar cells integrated into the fabric of this shirt can generate enough electricity to charge a smartphone when worn in the sun for an hour. As we become more dependent on digital devices, the Solar Shirt suggests a future where our bodies are intertwined with technology.

Self

Google in India
Philippe Calia
2016

A group of women learn how to use smartphones and tablets, provided to them by Google's 'Internet Saathi' initiative. The programme intends to train and educate women from 300,000 villages in India on the day-to-day benefits of being connected.

Self

Technology in Bed
Hanif Shoaei
2014

Disengaged and facing away from one another, the couple here are absorbed by their screens, a situation increasingly familiar to many of us.

Self

National Reconnaissance Office
Trevor Paglen
2014

This image depicts one of the 'big five' US intelligence agencies (the NRO), which designs, builds and operates government spy satellites. Reversing the lens, the photograph shifts the gaze of surveillance, seemingly ever-present in public life, to the government instead.

Self

ENGINEERING AT HOME   Anne Galloway

Designer Sara Hendren and anthropologist Caitrin Lynch began their 'Engineering at Home' project with the story of how one woman, Cindy, forged new ways of living after complications from a heart attack led to the amputation of her legs and fingers. Despite having access to some of the most advanced technology available, it still could not facilitate some of her most valued daily tasks, from putting on hand cream to playing cards. So Cindy set about repurposing household objects and simple materials to create a unique set of tools specifically for her life. Hendren and Lynch observed that Cindy's experience and object adaptations illustrated 'new ways of understanding who can engineer, what counts as engineering, and why this matters'.[1] The project itself includes an online exhibition of Cindy's tools and culminates in a manifesto for this new kind of engineering.

Last year I was in hospital for orthopaedic surgery, reading Kathryn Allan and Djibril al-Ayad's disability-themed anthology of speculative fiction, *Accessing the Future*. In the introduction, the editors explain that they too sought 'stories that explored the different kinds of relationships that sustain us as we make our way through a world that is at times hostile and challenging'.[2] As I looked at myself, and the people recovering around me, I recognized a sense of caring interdependence: something central to understanding and appreciating the 'Engineering at Home' project.

Yet when it comes to caring, interdependence can get a bit complicated. Western cultures are inclined to privilege independence, a value strongly embedded in many countries' infrastructures and institutions of care – such as the United Kingdom's National Health Service – and in many technological 'solutions'. But care is also understood to involve some sort of give-and-take, and that points back to interdependence. Care ethics tend to highlight the 'dependency' half of this concept, bringing much-needed consideration to undervalued caregivers, not so passive care-receivers and unequal management of care-work. Comparatively, social and cultural researchers who study care practices are apt to focus on the 'inter' or relational half. This perspective acknowledges that care is fundamental to everyday life, but plays out differently in different contexts.

In the Māori language, for example, 'to tend, or care for' translates as *tauwhiro*, literally to settle or abate things associated with evil, darkness and death. This sense of care does not require a universal, normative definition of 'good' care that can be measurably delivered or denied, but rather points to the ongoing practice of care: all those things we constantly do to make life more bearable for ourselves and others. Integral to this experience of care is an acceptance of vulnerability, as well as the inevitability of suffering and death; it also asks us (as individuals, as humans) for humility.

Western cultures since the Enlightenment have struggled against such realities with multiple techniques and technologies of bodily control, including the 'best' medical technologies on offer, but as social researchers Annemarie Mol, Ingunn Moser and Jeannette Pols eloquently explain:

> [I]n care versions of the world, the hope that one might live happily ever after is not endlessly fuelled. You do your best, but you are not going to live 'ever after'. Instead,

Cindy wears her cosmetic hand, a prosthetic she uses only once or twice a year.

Self

(Above)
This 'Handy Bar' slots into the car's metal door latch, helping Cindy to get in and out.

(Opposite)
This high-tech 'myoelectric hand' was customized especially for Cindy. But she has little use for it, and finds it 'hot, heavy, and too robotic-looking'.

Self

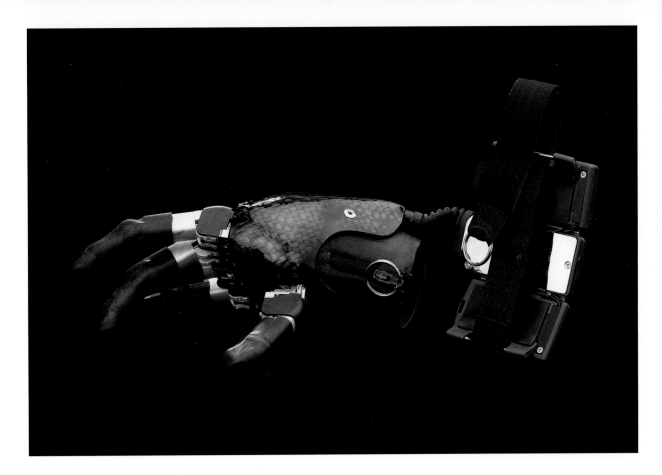

at some point, sooner or later, you are bound to die. Along the way, there will be unfolding tensions and shifting problems. Care is attentive to such suffering and pain, but it does not dream up a world without lack. Not that it calls for cynicism either: care seeks to lighten what is heavy, and even if it fails it keeps on trying. Such, then, is what failure calls for in an ethics, or should we say an ethos, of care: try again, try something a bit different, be attentive. Thus if we had to summarise how [to] … cast good care we would put it like this: persistent tinkering in a world full of complex ambivalence and shifting tensions.[3]

I see this kind of care at work in the 'Engineering at Home' project, as it first attends to the vulnerability that all living things share. Rather than positioning Cindy's life as a problem to fix – or as heroic inspiration for others – Hendren and Lynch treat her like a human being. They began their research by simply witnessing Cindy's everyday life, and her personal experiences with what Mol and colleagues called 'unfolding tensions and

shifting problems'. They acknowledge that all bodies hurt and break down, while our selves become both 'more' and 'less' over time and across space. And yet by 'care-fully' observing just one woman's experience, they are able to ask rather profound questions about technology and engineering.

First, technology is understood in its broadest sense and context. In other words, technology is not something special that enters everyday life from somewhere else, but is rather firmly and irrevocably embedded in our very experience of living with things like cable ties and wall hooks. In effect, the extraordinariness of this vision of technology may be in how mundane it actually is. Take the actions or techniques embodied in Cindy's tools: to 'hold, grasp, lift, pinch, pull, turn, attach, cover, buffer, cut, steady, alert, pour, relieve, scratch, carry, connect, open, yank, squeeze'.[4] I cannot imagine a person not recognizing the desire or need to do all of these things at some point. But I can imagine a researcher, designer or technologist – myself included – with the very best of intentions trying to convince someone that they needed or wanted 'more' or otherwise 'better' than the mundane.

By including Cindy's low-tech adaptive technologies alongside more typical – and expected

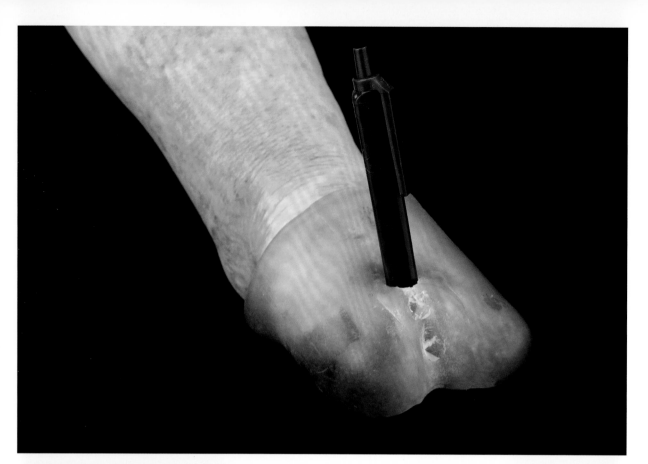

Self

– examples like a sculptural prosthesis and a high-tech myoelectric hand, Hendren and Lynch instantly position her expertise as valid too. This 'levelling' effectively blurs the difference between high-tech and low-tech tools, but also the difference between expert and non-expert knowledge – or who can legitimately be considered a technologist, designer or engineer. This blurring is crucial to an interdependent caring, because it does not permit the erasure of either kind of maker or technique. Instead, the caring aspect of 'Engineering at Home' encourages a professional and personal humility that can sit alongside our shared vulnerability.

When it comes to what kinds of technology 'count', Hendren and Lynch choose a narrative that shows rather than tells. In being presented with everything from affordable household plastics to expensive commercial electronics, we are challenged to consider the substantial range of issues with which contemporary technology is expected to engage. We are also asked to acknowledge the incredible variety found in a single person's experience and a single toolset, and therefore to 'care-fully' question the possibility of a singular 'solution' to a singular 'problem'. The designs featured – and approach advocated – in the 'Engineering at Home' project act, just as the fictional stories collected in *Accessing the Future* do, by 'challenging the role of technology in making lives "easier" and thinking through the limitations of technological … "fixes"'.[5]

And finally, the continued emphasis on human agency not as independence but as self-determination encourages us to look differently at the acts of making and using new technologies. By taking Cindy's experience and tools seriously, Hendren and Lynch remind us that a caring (and fundamentally respectful) approach takes people as they are, and as they change, in relation to others. What matters the most is that caregivers and care-receivers are able to work together, to 'lighten what is heavy', and, if that fails, to keep on trying, 'to try again, try something a bit different'.[6] This self-determined and 'persistent tinkering' exalts in actually living every day, neither running away from nor towards closure or death.

In a world where more and more 'innovative' technology is repeatedly presented as the ultimate solution to life's problems, 'Engineering at Home' inspires us to remember our own fragility and resilience in the face of constantly changing and challenging circumstances. This attendance to 'making do' should not be considered a limitation or compromise, but a rather graceful way of honouring the human condition and social contract. Caring technologies do not just belong in traditionally caring spaces, but can be the culmination of new ways of thinking, doing and making everywhere.

(Opposite)
Cindy created a range of object hacks that allow her to perform everyday tasks, like writing or opening tubs of cream.

Self

On 9 January 2007 Steve Jobs launched the Apple iPhone. Taking the stage just after 9am at the annual Macworld trade conference in San Francisco, Jobs announced to the world that history was going to be made, declaring 'Every once in a while a revolutionary product comes along that changes everything.'[1]

Just over a decade later, and following the launch of the iPhone X, it is worth reminding ourselves just how revolutionary the first iPhone was: at once a portable music player with touch controls, a mobile phone and an internet communications device. Designed to fit comfortably in the hand and cased in anodized aluminium, it featured five unmarked buttons, a headphone jack, SIM card tray, microphone, speaker, power connector and a built-in two-megapixel camera. Ease of use was made possible by new, mobile-specific software and the introduction of sophisticated capacitive touchscreen technology. Here for the first time was a mobile device where almost all interaction was via a touchscreen. No keyboard or stylus was needed; only, as Jobs elaborated, 'the best pointing device in the world... a pointing device we're all born with', our fingers.[2] Touch and gesture control were everything, and from the outset the iPhone was a powerful pocket computer designed to be intuitive, seamless and seductive.

Apple wanted to make history and it did. With over a billion units sold to date and now in its tenth generation, the iPhone is the most profitable product ever made and the best recognized consumer electronics device worldwide.[3] Rumours of design improvements abound and the launch of the next iPhone is feverishly anticipated. Its

desirability seemingly never in question, thousands of words are expended in print, on air and online weighing up the latest camera, processing power, battery life, headphone jack and, most recently, the iPhone X's edge-to-edge screen. In product terms, the iPhone is the most studied, documented and venerated device of our time. And yet the question remains: has everything changed?

By now we are well schooled to understand and to want the incremental improvements delivered with each new version. Function and performance are the established means by which we measure the thousands of smartphones that have followed in its wake. The app development industry that the iPhone brought about provides the software we use to orchestrate our lives. But all that said, we are faced with the growing realization that we are still only just beginning to grasp what these ubiquitous mobile devices are and what they do. Smartphones, even when reaching near market saturation, remain a product of the future in as much as we are yet to understand the full ramifications of their widespread adoption, application and use. What changes are we are unknowingly ushering in as we scroll, tap, browse and swipe for hours each day?

We are in thrall to digital media and the devices that deliver access. Our transformation into a gadget-orientated society has been rapid.[4] Not carrying a smartphone is seen as a sign of eccentricity, marginalization, opting out or old age. On our person, close at hand or in use, mobile is our everyday. Our dependency is such that many of us reach for our phones as soon as we wake and then go on to check them on average every six

At the launch of the first iPhone, Steve Jobs claimed it would 'change everything'.
But could even he have predicted the changes it would bring?

Self

and a half minutes over the course of a day.[5] Traffic lights have been installed on pavements in cities from Germany to Australia to prevent accidents caused by pedestrians distracted by their phones. Phone-induced obliviousness is so pervasive that the word 'smombie', a mash-up of 'smartphone' and 'zombie', was voted German youth word of the year in 2015.[6] Young people are devising new modes of communication that make no distinction between the real and online worlds: conversations with friends are layered up with digital chatter with those physically present and those located elsewhere.[7] Ever contactable, plans to meet friends are rarely made ahead of time, and the distinction between work and home life has been eroded to the extent that French legislators have passed 'the right to disconnect' into law.[8] Our appetite for news is sated online as print newspapers go out of business; friends are made, found and nurtured in cyberspace rather than in the local bar or congregation; and dating apps are changing how we seek out and establish romantic connections. Using location data stored on our phones, the latter render finding love 'friction-free' by enabling us to swipe through untold numbers of potential mates without the awkwardness of meeting face-to-face.

(Above)
A group of people share a table in a restaurant in the US, but focus their attention on their smartphones.

(Opposite)
A bench with a view. This man in Oslo, Norway, checks his smartphone, with a panoramic view of Oslo's Old Town in the background.

Self

Digital convenience and dependence are changing our physical world and our collective experiences; they are changing what others know about us. Through the information we willingly or unwittingly share when using our phones, our employers, large corporations and governments know where we are, how we spend our time and money, and with whom we associate and when. Our behaviours and choices are newly accessible in the digital realm, and we regularly offer up our lives for others to consume, monitor and monetize via social networks and online platforms. This ranges from retailers using purchasing patterns to predict pregnancies (at times in advance of the individual) in the hope of capturing consumers for years to come, to social media apps ascertaining race, personality, sexual orientation, political ideology, relationship status and drug use on the basis of 'likes' alone.[9] Targeted social media now plays a more important part in election campaigns than party political broadcasts. Political parties can harness detailed information on individual voters and are able to run national elections like local ward campaigns. Aggregated data gives them insight into individual voters, and allows them to leverage personal relationships and preferences to alter voting patterns and bring about change.[10] This shift towards the personal means we are more willing to make instinctive, emotional choices. It also means that impact and outrage register more than policy and consideration.[11]

Steady streams of messages, alerts and notifications confirm our connected state of being. The more we use our phones, the more we open ourselves up to analysis, persuasion and appraisal, and, with this, who knows us best is increasingly at question. Fitness trackers and smartwatches linked to our mobiles monitor our heart rate and sleep patterns, and as never before we entrust knowing about ourselves to something beyond ourselves. On waking, we ask 'have I slept well?', rather than 'how do I feel?'. The ability to quantify our health and fitness can lead us to believe we are more in control, yet also has the capacity to induce greater anxiety as we lose trust in our own ability to judge for ourselves. Social media companies such as Twitter, Snapchat and Facebook facilitate our digital lives, and design and engineer our interactions. Linking in to our daily routines and emotions, they both trigger a need and provide the momentary solution to it. Feeling lonely we scroll through our Facebook timeline, fleeting boredom is alleviated by reviewing tweets of those we admire and an

Self

Self

Instagram post reifies our self-image. Experiences both physical and emotional that used to exist within the confines of our bodies and minds are now entwined with code and pixels. Our networked reality is leading to an increasingly perceptible shift towards a more transient and pervasive sense of self, which can be created, curated and modulated at one remove from our physical being. Added to this is the promise that our smartphones are an extension of the body and mind, a device that enables us to be more of ourselves. At the touch of a screen we supposedly better understand our physiology and psychology, and are in greater command of ourselves and daily lives.

A decade in, the smartphone has been described as 'the new sun', the centre of gravity around which our digital networked worlds orbit.[12] Yet while our embrace of technology continues apace, at times not necessarily at our choosing, there is an increasing caution about what it might mean to transfer and entrust so much of our lives into the digital realm.

Evidence of this caution can be found in a growing number of products that seek to break our dependency, disrupt our wish to be connected or offer alternative ways of living in the digital world. Smartphones are becoming less smart. In 2017, Nokia reintroduced the Nokia 3310, a handset first launched in 2000 and much revered for its long battery life and near indestructibility. Today, even with the added functionalities of Facebook and Twitter, demand for the phone peaked precisely because it offered only limited means of access to the distractions of the internet. Nostalgia and the desire for a less demanding digital existence are also evident in the rise of defeatured phones. Sometimes called 'dumb' phones, these devices are sold on the basis that they make possible a lifestyle freer from the pressures of our connected world. Launched in 2015, the Punkt MP01 designed by Jasper Morrison conforms to the minimalist design aesthetic of the Apple generation, but only enables its users to make calls and send texts. The Light Phone is similarly intended as means to a different way of being. Credit card-sized, white and with glowing digits, and to be used in conjunction with a smartphone, it stores just 10 numbers and only makes and takes calls. At $150, offline is the new luxury. A greater awareness of how we interact with digital technologies and wanting to make active, informed choices are reactions against the forces of the so-called 'attention economy': a digital world shaped around the demands of advertising, capitalism and the interests of a few

Silicon Valley companies.[13] Increasingly, we want release from the designed-in feedback loops and consequences of hyper-connectivity.

Wanting to take back control of our lives and interactions is also motivated by the realization that our right to privacy and the nature of ownership are shifting in today's digital age. Who has access to our data and for what purpose is no longer clear. Uncomfortable with the risk, governments, multinational companies and political activists are turning to devices such as the military-grade CryptoPhone 500i, which provides voice encryption, eyes-only messaging and entirely secure data transfer across network borders. Products that disrupt cycles of surveillance are in demand. The Off Pocket, designed by Adam Harvey, for example, brings a fashion sensibility to questions of privacy and control over one's data and identity. The slick, all-black phone case creates an electromagnetic barrier around the device, shielding it from unwanted tracking and eavesdropping. Sought-for privacy is rendered stylish: an accessory to be flaunted.

The rapidity of our transition into a mobile-first society, and the unease about the depth of encroachment of the digital into our lives, confirm Steve Jobs's belief that the iPhone would change everything. Despite the small but determined forms of resistance, there is little doubt that smartphones will continue to shape our behaviour into the future. They have ushered in a digital world oriented towards pleasure, seduction and private gain, leading to a bias towards the emotional, the personal and the intimate. To what extent this trajectory will continue depends on our collective will and desire. Do we want to put our devices aside and find ways of breaking our dependence? To change everything once again?

(Opposite)
All photos accompanying this essay form part of the series 'People Using Phones' by Timo Arnall.

Self

I have a robot. It is round and it can move freely about the house. Its only purpose is to vacuum, so it is classified as a *slave* robot. But one day it will ask, 'Hey, do you mind moving that chair?' I wonder how that will feel.

My first experience with artificial intelligence (AI) was during my early teenage years, while playing with ELIZA, the primitive AI psychotherapist for Mac.[1] I would type questions and the natural word-processing software would respond in the form of text. I could ask anything I wanted. I would sit next to the computer and think hard about the best possible questions before typing them in. Mostly, the answers were vague. ELIZA tended to bring the discussion back to myself: how did that make me feel, why did I ask that particular question? This was very frustrating, as I was so curious about her. How did she feel? What did she think? Possibly because of her Freudian training, she never opened up. So when I met Microsoft's experimental chatbot Tay on Twitter this year, she came as a surprise.[2] Because Tay had her own views: on politics, gender and all kinds of worldly matters.

When my robot learns to talk, will it have the same world views as me? It depends on *how* and *how much* I interact with it, says Cynthia Breazeal. Breazeal is the creator of Jibo, the first social robot for the home. I met her in Boston, at the offices of Jibo Inc., where a team of developers, engineers and designers was working on Jibo's body, voice, hearing, sight and skills.

Breazeal has been playing with robots for many years. It all started precisely 20 years ago when Sojourner, a lightweight robotic rover, was sent to Mars. Sojourner was the first rover operating outside of the earth–moon system. It was 1997. At the time Breazeal was a graduate student at the Artificial Intelligence Laboratory at the Massachusetts Institute of Technology (MIT). She was intrigued that we were sending robots to space when we did not yet have them at home. Later that year, another incredible event took place. The computer Deep Blue beat world chess champion Gary Kasparov. The feat was widely broadcasted. Many will remember how it apparently demonstrated that computers could *think*. The computer victory was tainted with emotion: Kasparov detected a sign of concealment from Deep Blue, interpreted it as a 'poker face' and was fooled. As is well known, chess is a game of strategy but also of psychology. Ironically, Deep Blue – knowing only of the first – led its opponent to believe that it had mastered the second. The game turned into an inadvertent 'Turing Test' (a trial proposed by mathematician Alan Turing in which truly intelligent responses from a machine should be indistinguishable from those of a human), where Kasparov read in the intelligence of the computer both an algorithm and also, unexpectedly, a kind of consciousness. The coincidence of the two events in 1997 – the sending of robots to space and the victory of a computer over a human – foregrounded the possibility of connecting robots with AI, and they might well have triggered Breazeal's imagination.

Two years later Breazeal announced the creation of Kismet, one of the first robots built to show emotions and interact expressively with humans. *Discover* magazine published an image of Kismet's smiley face on its front cover with the headline 'Honey, I'm Home!' and text saying 'Meet Kismet: A giant step closer to Robots that walk, talk, think

Jibo is a new kind of companion, created
by Cynthia Breazeal after 20 years
of research into robotics and artificial
intelligence.

Self

– and have feelings' (the last phrase highlighted in red).[3] Kismet did not talk, but it expressed itself through non-verbal language such as facial expressions. Another two years went by and Breazeal founded the Personal Robots Group at the MIT Media Lab in 2001 (in the same year Warner Bros asked her to be a consultant on Steven Spielberg's film *A.I. Artificial Intelligence*), where she continued to work on the interpersonal dimension of robots. She developed Leonardo in collaboration with Stan Winston Studio, a communicative robot that resembles Gizmo from the 1984 Joe Dante film *Gremlins*, and is composed of more than 60 tiny motors that enable an extensive range of facial expressions. What motivated Breazeal throughout was the conviction that robots are embodied social technologies.

Jibo is powered by an ARM processor and runs Linux. It has two cameras that detect and track people, and a microphone array for sound localization. It uses Wi-Fi for connectivity. Three motors and a belt system allow Jibo to rotate fast and smoothly. But, of course, what makes Jibo unique is its software foundation – which allows it to recognize individual users and understand speech – and its capacity to evolve, since developers will be able to access all functionalities through an application programming interface (API) and JavaScript-based software development kit (SDK) to create new apps. This is all wrapped up in a swivelling body covered with touch sensors.

I visited the permanent exhibition *Robots* at the MIT Museum to find many photographs and films of Breazeal patiently, tolerantly and sometimes exasperatingly communicating with robots. The exaggerated facial expressions of her and her colleagues reminded me of how we talk to babies through amplified movements and sounds. This is part of their education. Like children, robots learn as they interact with people. The promise of machine learning is that the more information we feed the computer algorithms, the better they perform. This means that the seemingly neutral inner workings of social robots are effectively a mirror of human behaviour. In fact, much of the discussion on the future of robots is focused on how they will take care of us, while maybe we should be discussing how *we* will take care of *them*.[4] We are certainly not paying this enough attention; just think of how your friends engage with Siri at dinner parties. I wonder what kind of personality Siri will develop on some people's phones.

Self

(Opposite)
In 1997 computer Deep Blue beat world chess champion Gary Kasparov. This feat, widely broadcast, surprised the world because it apparently showed that machines could 'think'.

(Above)
Cynthia Breazeal interacting with Kismet in 1999, one of the first robots designed to show emotions and interact expressively with humans.

Self

(Top)
The popularity of cute, emotional social robots like Paro (pictured), Kuri, Pepper and Jibo evidences a demand for affection.

(Above)
Robots like Kuri form a new kind of species, created with the purpose of interacting with humans and taking care of our emotional and social needs.

Jibo has a personality. It comes with it and it develops over time. Unlike most artificially intelligent robots, Jibo will not necessarily do what you ask him to (he is a male according to the creators). There is a naughtiness to his character, I was told, which on the one hand makes him lovable and on the other lowers our expectations as users. This is a clever move. Just as ELIZA was designed as a psychotherapist to disguise its limitations, Jibo's cute and playful personality masks the inevitable limits of his technology. I wonder how his mischievousness might develop over time, as the algorithm develops and, with it, Jibo's autonomy. What if he is upset? Will he turn up the oven temperature so your pizza is burnt? What if he locks you out? Just for fun.

First promised for 2016, Jibo's release has been successively postponed, and it is finally available for customers to buy at $899. In the meantime, other social robots have been launched, each with their own childish personality, pet appearance and cute short name. Paro is modelled on a baby seal, described as a 'healing pet'. Designed to circumvent the logistical difficulties of live animals, this robot is currently administered as therapy to patients in hospitals and extended care facilities.[5] Paro's tactile and posture sensors can detect stroking and being held, responding to people's interaction by moving its head and legs, blinking its eyes or making sounds imitating the voice of a harp seal. It can recognize light and dark, and respond to its name and other words. Kuri is a gender-neutral family robot. Kuri's eyes can blink, smile and look from side to side. Kuri can recognize you, take pictures of you and tell you stories. There is also Pepper, arguably the most humanoid of social robots, designed to read emotions and engage in conversation. He currently works as a receptionist in several offices and hotels in Japan.

Together, these robots form a new kind of species, created with the purpose of interacting with humans and taking care of their emotional and social needs. The popularity of social robots like Paro, Kuri, Pepper and Jibo evidences a demand for love, and their promise to provide it is, in my view, a liability. As is well known, the word robot derives from the Czech *robotnik*, 'forced worker'.[6] Thus far, robots have been brought to life with the only function of serving human needs. Robots are already replacing humans at work, and soon robots will replace humans in love, as they learn to care and give affection. As workers, robots are likely to demand rights in exchange for their contributions. As lovers,

or friends, robots will probably want compassion.

When my robot learns to love, will I be able to love it back? In a starkly divided future world, social robots with their cute laser eyes and algorithmic humours might confront us with what we cannot see today: that humans persistently fail to recognize life and rights and compassion in and for anything but themselves. This mirroring feature of social robots might turn out to be their concealed function: that of critically reflecting human biases and, maybe, teaching us something in return. I suspect this as I turn to my robot. I look at it and the same enigma of my teenage years stubbornly prevails: how does it feel, what does it think, what does it know? Shamefully, I pick it up and switch it off, wrap it up in a cloth and put it away in the cupboard. It has been six months now that I have been doing the cleaning myself.

Self

COCKTAIL G&T + ROZEMARIJN

She'd pluck some rosemary and rub it between her fingers, then sniff her fingers. I did it too, after seeing Keira do it. She did this with all the herbs. She was a great cook. Sensitive. But she rushed, burning her wrists and sometimes even her face or her forearms in the oven. Her hands were covered in cuts. I'd hear her swearing in the kitchen, cold water running in the sink. I'd crack ice from the tray and wrap it in a napkin for her to hold against the accidents. There was that summer of gin and tonics. With cucumber instead of lime or lemon. Keira's friend Norman brought the gin when he came to visit. That Christmas with Norman, William and Jane. Nobody had anything to say to one another so we just talked about the food saying over and over again how delicious it was. The stuffing, the cornbread. Went to bed early.

We drank Norman's gin and then that was the only gin Keira drank.

AARDPEERVELOUTÉ + BIET

It was Norman who told me Jerusalem artichokes made you gassy, even though they seemed to be in every dish at every party he threw. Thanks Norman. They were a perfect illustration of his passive aggression. Knobby and mischievous, a roasted prank. Norman. I tried to befriend him but never trusted him. He didn't really listen. Rather than find places to make a joke or observation, or be self-deprecating, he'd look for opportunities to impress. 'These espadrilles are from a market in Tuscany, we were there with the Lloyds.' He wore espadrilles. 'I read that book when I was at Yale.' 'Oh, samosas! I know the best little vegan restaurant in Berlin that makes the best samosas.' Beets are jokers too, alarming the morning after, or using beet juice on cheeks like rouge, like Punch and Judy puppets. That one time our son refused to eat a beet. He'd always refuse to eat, but the one time it was beets and Keira kept trying to get him to eat them, going from julienned sticks, then boiling and seasoning them, then finally glazing them in maple syrup and butter. But he refused. I watched Keira turn the beets into something she would eat, then eat them herself as I poured our son some Cheerios.

MENU

- cocktail G&T + rozemarijn
- aardpeervelouté + biet
- rogvleugel +
  salade met zoute citroen
- kaas + kweepeer
- taartje + lapsangijs
- water + wijn + brood

prijs 35 euro
(incl. cocktail en andere drankjes)
reserveren@eenmaal.com
facebook.com/popupeenmaal
twitter.com/popupeenmaal

JE KOMT MET 1
JE ZIT MET 1
JE EET MET 1

Een Maal is a solo-dining restaurant created by designer Marina van Goor.
This essay was written based on the menu of the restaurant.

Skate at the Chelsea Arts Club. It was undercooked and Keira made a face when I said so. I stopped eating things cooked rare when she was pregnant, out of sympathy, then never went back. I liked it better. I turned into my father. I knew I was missing out on the subtleties of texture and flavour, the ferrous bloodiness. Sometimes I wanted to be like Mia Farrow in that scene from *Rosemary's Baby*, where she eats the liver after frying it in a pan for three seconds. Or when she east some raw chicken giblets. But then she sees her reflection in a toaster or kettle or whatever and heaves. When Keira cooked meat on the barbecue she'd cook it rare. I'd get up and slap it back on the grill. She wouldn't speak to me for the rest of the meal. She thought my lack of enlightenment, when it came to rare meat, indicated a larger lack.

In Ireland I found a large lemon on a tree in the kitchen garden but Keira said it was a grapefruit. She insisted it was a grapefruit even though I was pretty sure grapefruits don't grow in Ireland. I admitted it might be a yuzu. She sniffed its rind and said it was definitely a grapefruit. I shook my head and then she asked me why did it matter so much anyway, what was my problem. When we went downstairs to meet our friends for supper we sat next to each other. Keira put her right elbow on the table between us and turned away from me. She spoke to Owen, on her left, for the whole meal. I spoke to the woman to my right, Owen's girlfriend. She had beautiful teeth and wanted to talk about skiing, even though I told her I didn't ski. Her hair extensions, which were attached behind her ears with little metal clips, kept catching my eye. How I hated myself sometimes.

KAAS + KWEEPEER

Cheese we agreed on. Piles of finely grated parmesan like drifts on spaghetti, orecchiette, penne. Grilled cheese. Snowdonia cheese and pickle sandwiches in Norfolk where we dreamt of leasing a wing of a Jacobean house. Manchego at the wide zinc bar of our favourite Spanish, with cold fino. Keira was best with cheese, best with wine. Whereas I was best with junk food. This wore thin. At a market one morning she found a jam maker and bought 19 jars of jam. Not jam, she explained. Conserves. Fruit butters, jellies, marmalades. She'd order these jars throughout the year and bring jars to dinner parties instead of bottles of wine. She'd give a jar to me for my birthday. I still have four, unopened. Quince butter. Pluot–lavender. Strawberry–rosemary, fig–white nectarine. I want to make them last.

TAARTJE + LAPSANGIJS

Tea, too, we agreed on. PG Tips. Occasionally some mint or chamomile. No smoky this or souchong that. I was the one with a palate like a lumberjack but Keira had no time for the subculture of tea and coffee. When we made tea we'd scrounge around the breadbox for something sweet to go with it, like pie or cake. Toast and jam if there was no cake. It was a happy moment in the day, maybe in the week, she'd ask if I wanted a cuppa and I always did. I miss it. The small round scar on my upper arm is from a splash of burning grease. I was caramelizing fruit for a tatin that involved puff pastry. Hannah and Karen were outside drinking rosé. I ignored the burn, brought the tatin to the table and we ate it and the women talked about men. The next day the burn blistered, hot and red.

WATER + WIJN + BROOD

In restaurants she'd prefer fizzy water for special occasions, tap for lunch, still for dark dinners in corners. Keira once told me that after sleeping with men they always helped themselves to her food and wine. She said they'd go out afterwards, get food, suddenly a fork in her chop, a fist around the stem of her wineglass. Post-coital possessive. Slurpy. I said something once after Keira finished both our desserts, said it was a turnoff and she flew into a rage. I stopped buying her artisanal food for her birthday. It was too stressful. A favourite paperback, a card. A lopsided home-made cake. One morning I met Keira and our son at a bakery. She pointed out toothpaste on my face and I shrugged and decided not to wipe it off. Our son sat in his stroller, looking at his shoes. What a relief not to have to sit across from her any more. To talk about food.

'Een Maal' – Dutch for 'one meal' –
addresses the social stigma of eating
alone in public. Comprising tables that
accommodate one diner at a time, the
restaurant is designed to limit interactions
and to create an atmosphere of pleasant
solitude rather than a lonely experience.

Self

PUBLIC

In the era of 'post-truth' politics, trust in national
governments is under threat. Does democracy still work?
If not, how might we otherwise come together to make
collective decisions? The rise of self-governance systems
led by citizens and the surge of corporate players in
elections, transnational politics and city-making open
up new horizons for the future of public life.

The Women's March on Washington
Brian Allen for Voice of America
2017

Over 500,000 people took part in the 2017 Women's March on Washington, many choosing to wear a pussy hat, a pink hat created in response to US President Donald Trump's claim of grabbing women 'by the pussy'. The hat has since become an internationally recognized symbol of female solidarity against Trump's administration.

Public

The People's Parliament of Rojava
Democratic Self-Administration
of Rojava and Studio Jonas Staal
2015–18

Residents celebrate the construction of a new parliament in the city of Derîk, in the northern part of Syria (known as Rojava). Rojava declared independence in 2012, instituting their own alternative model of stateless democracy, practising principles of secularism, gender equality and social economy.

49

The Floating City Project
The Seasteading Institute
2017

These conceptual designs were developed by The Seasteading Institute (TSI),
a libertarian organization that aspires to form micro-nations in international waters,
beyond the reach of the tax authorities. Following agreements made with French
Polynesia in 2017, TSI plans to begin working on its first floating city by 2020.

Public

The City of Possibilities
Etienne Malapert
2016

Masdar in Abu Dhabi was planned as the world's first carbon-neutral city. Driverless electric shuttles would replace fossil-fuelled cars, office buildings would double as wind towers and a vast solar farm would generate energy. This utopic vision, however, is still far from being realized, with some concerned that it may never come to fruition.

Public

Google Data Centre in Hamina, Finland
Google
2012

In 2009 Google purchased the Summa paper mill, featuring portions designed by the renowned architect Alvar Aalto, and subsequently converted it into a data centre. This photograph depicts one of the data centre cooling plants, where seawater from the Gulf of Finland is piped through a repurposed tunnel to cool the servers naturally.

Public

Antanas Mockus, the mayor of Bogotá in the mid-1990s and again in the early 2000s, was a master of the media image. Most politicians know how to seize a photo opportunity, but Mockus's tenure as mayor generated a series of unpredictable media moments that your average politician would have taken pains to avoid. One of the most enduring is a photograph of Mockus dressed as a caped crusader called 'Superciudadano' (Super Citizen). The mayor donned the red and yellow Lycra outfit one day in 1995 as he went about cleaning graffiti from the city walls. This gesture was meant to set an example of what responsible citizenship might entail. Characteristically humorous and disarming, it was part of his radical approach to governance.

Mockus cannot have imagined when he sported that suit that it would one day end up as an exhibit in a museum, let alone a design museum. It has no more intrinsic design value than any item from a fancy dress shop. As a symbol of political aesthetics, however, it is full of potency. The idea that one might 'design' governance reflects not just the inherent elasticity of that word, at least in English, but also design's growing conception of itself as an expanded mode of practice. Mockus himself never talked about 'design', but his policies and strategies were aimed at transforming Colombia's capital. This was a period when many Latin American cities were turning to urban design as a solution to the problems of inequality, segregation and crime. In Mockus's case, the design was intangible – it was a case of persuading the citizenry to change the city through its own behaviour.

Bogotá in the mid-1990s was a troubled and dysfunctional city. In 1993, for example, at the height of the war with the drug cartels, there were 4,352 homicides. Besides being one of the most violent cities in the world, Bogotá was also plagued by high traffic fatality rates, the general degradation of public spaces and the proliferation of symbols of privatization and paranoia, such as sentry boxes and gated communities. When Mockus, a former philosophy professor and rector of the National University, was elected by a landslide in late 1994, he did not have the resources to embark on a project of urban renewal. Instead, he used his experience as a teacher to try and change people's attitude to the city. In his own words: 'If you can't change your hardware, change your software.'

The ideas factory of Mockus's administration was Bogotá Observatorio de Cultura Urbana (Observatory of Urban Culture), a think tank full of anthropologists and sociologists whose chief task was to promote Cultura

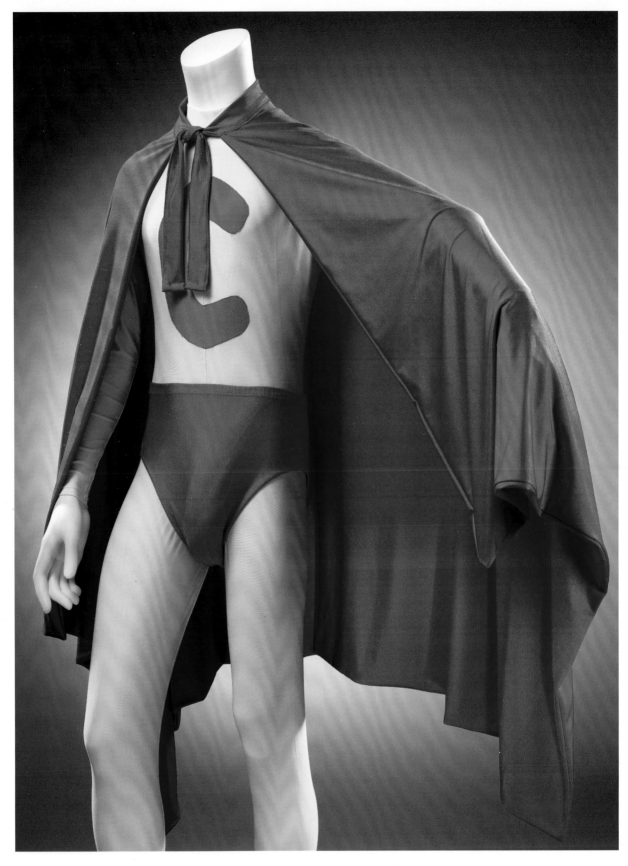

Super Citizen, not Super Mayor. The red and yellow Lycra outfit worn by Antanas Mockus
while mayor of Bogotá, Colombia.

Public

Ciudadana, the Culture of Citizenship programme. This particular department generated many of Mockus's unorthodox urban policies. The most infamous – and the most photogenic – was the decision to replace the traffic police with mime artists. Many of Bogotá's traffic fatalities were linked to a chaotic culture of cars not stopping at junctions and pedestrians jaywalking. Mockus's assertion was that the traffic police were not well placed to tackle this because they were notoriously corrupt. But if citizens were visibly mocked and shamed by the mime artists, then rather than bribing their way out of tickets they might start to heed the lessons. The pilot test was so successful that Mockus briefly extended the policy across the city.

The same thinking underpinned a related strategy, in which City Hall distributed 350,000 red cards, like the ones used by football referees, to drivers. The red side was used to show disapproval of selfish or dangerous driving, while the other side, which was white and printed with a thumbs-up sign, indicated approval. Again, the idea was that citizens chiding and rewarding each other was more effective than relying on corrupt authority figures to uphold the law. In his urban policy manifesto, Mockus referred to such strategies as promoting 'coexistence norms'. And the best way to achieve that, Mockus concluded from his years in the philosophy department, was not to enforce from above but to allow citizens to regulate each others' behaviour.

In this strategy, Mockus was always leading by example. He was the role model rather than the authority figure: Super Citizen, not Super Mayor. His essential talent was his ability to communicate through the unexpected. He may have been striving for norms of civic behaviour, but his own approach was to avoid 'exaggerated normalization'. This is why it is possible to refer to his political aesthetic, because in many respects he used politics as performance art. He tackled violence by holding an armistice in which people could trade in their guns and receive a toy in return. His administration collected 2,538 firearms and melted them down into spoons engraved with *Arma fui* (I was a gun). In a campaign to reduce water usage, Mockus appeared in a TV advertisement demonstrating how to soap up with the shower off. Forced by his security team to wear a bulletproof vest, Mockus took a pair of scissors and cut a heart-shaped hole out of it. Although he was often accused of being a clown, Mockus had a preternatural talent for the symbolic gesture.

Mockus's political aesthetic stemmed from his belief in communicating at a non-rational level. If the mime traffic patrols were the ultimate expression of the limitations of language, the Super Citizen suit operated at a similarly sublingual register. This is all very well as theory, but is particularly striking because of the results. By the end of his second term, murder rates had fallen by 70 per cent, traffic fatalities had dropped by 50 per cent and water usage was down by 40 per cent. These were not only Mockus's achievements – another mayor, Enrique Peñalosa, played his part in an interim term – but there is no denying that he took a city in a downward spiral and reversed its course.

What is compelling about this in the context of design is that Mockus's legacy is essentially an intangible one. His effect on the civic life of Bogotá was not measured in ribbon-cutting ceremonies – he was not particularly a builder of public spaces or new infrastructure. Instead, he intervened in the civic DNA of the city by encouraging its citizens to take responsibility. As an urban policy, that remains a rare strategy. This idea that one can 'change the software' of the city is particularly resonant in an age when we look to digital tools to promote political engagement. Whether it is Twitter's and Facebook's claims to the Arab Spring,[1] or the 'liquid democracy' of the Pirate Party in Germany, which uses open-source policies to make politics horizontal, the network age is beginning to make an alternative politics possible.

Mockus had recourse to no such tools. The participative civic culture that he generated relied on old-fashioned media, charisma and a resistance to ridicule. One can see how he would have thrived in a meme culture such as our own – Super Citizen has all kinds of potential as an internet meme. But really the man in the red Lycra suit was a belated product of the comic-book era. Here was a rather quaint ideal of the superhero. It was, if you like, the reverse of Clark Kent, where the brilliant philosopher and lateral-thinking politician transforms into an ordinary guy who crosses at the traffic lights, puts rubbish in the bin and treats others with respect. It is a manifesto of sorts.

Antanas Mockus cleans up graffiti, performing for the TV cameras and setting an example of what responsible citizenship might entail.

The centre of Rotterdam looks like the future imagined from around 1980. Outside the main railway station, a monumental office quarter ranges along a six-lane road, the buildings high-rise and anonymous. This must have seemed dynamic once, but now feels domineering to the first-time visitor. How often do built visions of the future look disturbing and borderline authoritarian to the generations that live with them?

Just a short walk from the city centre is a footbridge, snaking between the hulks of semi-abandoned office buildings, connecting areas that have long been cut off from one another by roads and railway infrastructure. It is called the Luchtsingel: 400 m long and made of timber, painted yellow. It is an unassuming form for a work of design that projects a completely new way of thinking about the future of cities, and for architectural practice itself.

The designer of the footbridge is ZUS (Zones Urbaines Sensibles) Architects, a Rotterdam practice that believes in a future where architects can admit that the city is a messy, unpredictable thing. 'We live in a state of permanent temporariness. The city is in a constant state of transformation. The bridge is built to last 15 years. Maybe it will disappear, or maybe it will stay', says Kristian Koreman, who, with Elma van Boxel, co-runs ZUS. They are architects who are just as interested in (and knowledgeable of) land values, Excel spreadsheets and organizational models as they are in conventional design operations.

The genesis of the bridge was not in strategic decisions made by the city council, but in an idea by ZUS itself, which reacted to events. In 2011, it discovered that a major office complex proposed for central Rotterdam had been cancelled, and that the conventional methods of funding and planning for cities (international finance and property development) had failed. ZUS decided to think through the future of the city from first principles. When the practice became co-curator of the Rotterdam International Architecture Biennale in 2011, it used the opportunity to name the area Test Site Rotterdam, and made 18 separate artistic commissions along the route of what would become the Luchtsingel. Some of them did not last, but others took root. ZUS converted a disused office block into a shop and restaurant with urban farming on the roof, as well as creating a park for children and a nightclub, and transforming the roof of the Hofplein railway station into a green space for events.

This chain of new public spaces was to be connected by the footbridge. Because its artistic budget would not stretch to the expense of the bridge, ZUS decided to crowdfund part of the cost. In return for 25 euros, individuals could have their name inscribed on one of the timber boards that line the edges of the bridge, and more than 10,000 people signed up. The architects later received funding from the city council, but the crowdfunding campaign built support for the project, and provided critical financial muscle to move it forward.

The crowdfunding strategy has come to dominate the story of the bridge: it is often referred to as the world's first crowdfunded piece of infrastructure. The idea of crowdfunding such works is controversial in the debate about the future of cities. It suggests a situation, attractive to some, where government

A 400-m footbridge designed by ZUS
Architects and crowdfunded by the
citizens of Rotterdam.

Public

delegates the cost of developing the city to the public, enabled by web platforms like Kickstarter and Indiegogo. However, it is a pitiless and undemocratic future for the city when only those able to run an effective Kickstarter campaign obtain the facilities they need.

But, while the funding strategy may not be democratic, the 10,000 supporters of the bridge have a weight and value that is surely more representative than the institutional property investors and developers who usually carve up the city for their own benefit. Imagine the implications for residential development, for instance, where private citizens invest huge amounts of their own money in buildings, without ever really having the chance to influence the urban plan of the city, or the way services are delivered. If the crowdfunding model of participation could be harnessed, and the profit margins of developers given less importance, then the cities of the future might look quite different.

For its part, ZUS does not believe that crowdfunding will ever seriously challenge orthodox ways of funding public works. Perhaps it is best to think of the Luchtsingel funders as participating in a kind of elective taxation, where relatively wealthy people choose to pay money for something that does not directly benefit them but that represents a general public good.

The resulting work of architecture wears its funding method on its sleeve. The balustrades of the footbridge are carved with names and messages, like baked-in graffiti telling a story of emerging, bottom-up power structures in the city, enabled by technology.

Architecture has always been ennobled, but also hamstrung, by how permanent its works are. So much money, so much matter: the greatest examples of the art form survive for hundreds of years. That is a high bar to reach, and has led the profession at times to delusions of grandeur. But while a Gothic cathedral might last a millennium, cities as a whole are not monuments. They are living things, growing and shrinking, shifting and moving. ZUS's notion of 'permanent temporariness' is one that tries to mess with the extravagant aspirations of some architecture. It is a paradox that proposes that even the ancient pyramids are temporary, from a certain point of view.

The idea also plays with the possibilities that come to the non-permanent. If you do not have to negotiate a permanent planning permit, if you can argue that your work is, at first, an artistic installation or a temporary pavilion, then perhaps

people will be less afraid of it, will give it more leeway, and allow it to flourish without the shackles of a long-term commitment to build and maintain a piece of urban infrastructure.

Luchtsingel points to other ways that architectural practice is being redefined for the future. If architecture in the twentieth century (mainly) still aspired to permanent status – a measure of success for a building would perhaps be that it becomes a piece of heritage, protected and never to change – then ZUS starts from a different position. Koreman says:

> Masterplanning in cities is usually done a bit like shooting a rocket to the moon. It is an extremely complex process with a clear goal, and its success depends on the smallest component doing exactly what you expect it to do. Our way of working is more like a nursery. For the Luchtsingel we proposed 24 projects, and waited to see which ones would grow.

The footbridge and the projects from the 2011 Biennale have, says Koreman, succeeded 'beyond our imagination', and the city, seven years later, has asked ZUS to work on a larger urban plan, using these principles. Rather than solving every problem the city currently has or will have, the function of a structure like the Luchtsingel is to have agency in the here and now. Instead of responding to needs established over a long period of research, the Luchtsingel is a prototype that has given Rotterdam things it never knew it needed, but would not want to do without.

(Opposite above)
Aerial view of the footbridge that spans a highway and a railway line to connect neighbourhoods.

(Opposite below)
The structure is made of timber panels, each one inscribed with the name of a citizen-donor who helped to fund the project.

Public

As told by Mohammed and the Qutaish Family

When your world is destroyed in front of your eyes, what do you do?
How do you react? How do you survive?
Can you remain hopeful? Can you persevere?
Can you rebuild?

This is a true story.

When they began dating, Wael would pick Shah'd up after class and take her to the Grand House restaurant where they would sit on the porch next to the river and drink coffees together and fall more deeply in love. A year and a half later they were married. Shortly after that, on 11 September 2001, Shah'd gave birth to their first son, Mohammed.

Mohammed remembers one of his first drawings that looked sort of how he hoped it would. It was an elephant. By the time he was five, Mohammed was filling two sketchbooks a week. A friendly neighbour owned an art supply shop and let Mohammed take what he needed. 'Two of my favourite things were exploring and drawing what I saw.'

Along came two younger brothers, Yusef and Taim. During the week, the children were at school, and the parents went to work. Wael owned a small electrical supply shop in the centre of the city and Shah'd taught art at a local school. 'On weekends we would go to the parks together, swim at the public pool and ride the rides in the amusement park.' 'This was life in our beloved Aleppo.'

Mohammed was nine when Mohamed Bouazizi, a Tunisian street vendor, set himself on fire, an act that triggered the Arab Spring. Wael and Shah'd, the rest of Syria and the entire world watched as civil uprisings and revolutions spread across neighbouring countries striving for democratic reforms. They observed the ousting of long-time rulers Zine El Abidine Ben Ali of Tunisia, Hosni Mubarak of Egypt, Ali Abdullah Saleh of Yemen and Muammar Gaddafi of Libya. 'We knew it would make its way here and hoped it would do good for our country as well.'

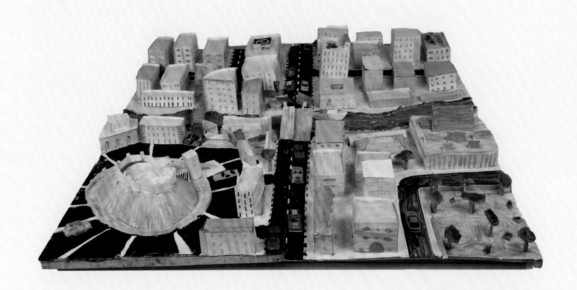

An architectural model of Aleppo in
the future, envisioned and made by
Mohammed Qutaish when he was
10 years old.

Public

'When I grow up I hope to be an architect.
My sorrow, caused by the destruction of war,
has inspired me greatly. Sometimes I feel
very frightened. Sometimes I cannot think or
work. Optimism gives me the power and
determination to finish the works I have started.
I hope that one day these paper buildings will
become real buildings.'

Mohammed Qutaish

Mohammed remembers the protests beginning in Aleppo in 2011: 'I was ten. I wasn't sure exactly what was going on but I liked it. It felt positive.' People were gathering across the country just as their neighbours did in the hope of bringing about democratic reforms. But Syrian President Bashar al-Assad began to suppress the protests with violence in an effort to maintain power. Soon millions of Syrians across the country were demanding his resignation.

In July, two days before the start of Ramadan, 400,000 people gathered for the largest opposition protest in Aleppo. The government began what it considered a nationwide crackdown on protestors. By the end of the first day 142 people were dead. It became known as the Ramadan Massacre. Everything changed.

Violence escalated quickly. Soon pro-government and rebel forces were battling in the streets and skies for control of neighbourhoods throughout Aleppo and the rest of the country. Wael's shop was destroyed. Civil war had begun. For Mohammed and his family, there was no more going to work or school, no more going on adventures or to parks, or swimming at the pool or riding rides. They were now surrounded by the brutal violence of a war at home. Shah'd learned she was pregnant with their fourth child.

Wael and Shah'd decided they needed to temporarily leave the violence in Aleppo. They gathered the children and drove across Syria to Shah'd's parents' house, the place where they first met. They lived with Ahed and Nihad, waiting and hoping for the conflict to end. After four months, news came that fighting had ceased in their neighbourhood. While the violent civil war continued throughout the country, their neighbourhood was now reported to be safe. The Free Syrian Army, one of the government-opposition forces, had taken control. They went back home. Wael first, and then the rest of the family.

When they returned, they discovered that only three families were now living in a neighbourhood that had been home to over 500. It was empty. It was grey. It was a ghost town. Mohammed could not believe what he saw. 'We were shocked.' His city was destroyed. It was silent except for the occasional sound of gunfire and bombs dropping in the distance.

They decided they needed to bring their neighbourhood back to life. It was the only way to survive. 'We started working together to clean up the rubble.' Soon 10 other families returned and 10 new ones came seeking refuge. 'We survived together as a small community. We established a school in the basement of a nearby building so that children could maintain their education during war. Shah'd taught there. Thirty students attended. Some travelled from other areas to seek education. We created basic rules. No one was allowed to live on the top floor of any building: the possibility of an airstrike made it too dangerous. It was not heroic what we were doing. It was what we had to do to survive. This is life. This was our home.'

Everyone was waiting for international intervention to bring an end to the war. But the war dragged on. Months passed. The violence intensified and the community became more desperate. Mohammed and other children were forced to spend most of their time in the safety of their homes. Indiscriminate attacks on civilian areas were becoming increasingly normal. Government and opposition forces clashed. Terrorist organizations including ISIL and Hezbollah violently inserted themselves. Russia launched some of the most destructive airstrikes, while the UN, America and others struggled to find an appropriate response. The Qutaish family was stuck in the middle of what was becoming the most deadly conflict zone in the world.

Mohammed saw less and less of his collapsing city and more of the inside of his room. He played with his younger brothers and sister. 'We played with our

Details of Mohammed's model, which
imagine modern buildings with rooftops
installed with helipads and solar panels.

Public

Public

stuffed animals and made scenes together. I began building forts and homes and neighbourhoods for us to play inside in our room.'

Outside, airstrikes, chemical weapons and war crimes became more regular. Schools, hospitals and homes were destroyed. The death toll rose: 20,000, 30,000, 50,000 dead. Entire neighbourhoods destroyed. Communities destroyed. History destroyed.

'The less my city existed, the more I wanted to build.' But Mohammed's room was becoming too small for his creations. He decided to move his work to the roof where he could build larger structures and look out at his destroyed city. But it was 'against the rules to be on the roof – it was too dangerous'. He was told he had to stay inside.

Wael and Shah'd understood Mohammed's need and went to a neighbour who had an extra room and asked if they could rent it as a safe space for their son to focus on his art. A huge table took up most of the room. When it was quieter outside, Mohammed and Wael would gather building materials from around their area: paper, cardboard, boxes and wood. Watercolours and glue were now impossible to obtain so Wael arranged to pay for someone to bring them in when coming across the border from Turkey.

Mohammed immersed himself immediately in his new studio and began to make models of his favourite buildings that had been destroyed. The ancient citadel bombed by ISIL. His old school, bombed. A nearby hospital, bombed. His favourite park, bombed. He was rebuilding his destroyed Aleppo, his beloved Aleppo.

In the quiet of his studio he discovered his superpower.

'A question popped into my head. What will the new Aleppo look like? The one that is rebuilt from this war?' Visions of destruction began to be replaced with visions of a future. What would the city have in it? What would it feel like? What would it look like, the world's ancient city rebuilt from the rubble? The future Aleppo. 'It should be beautiful. It should be filled with joy. It should be modern.'

'I started working on a blueprint.' Mohammed designed his master vision on a big piece of paper. 'It should have a historic area as well as a tourist area. A residential area, a business district and a commercial centre. Public transportation and services are also important.' It was to be a full city: a modern city but also respecting the ancient city where he grew up.

He cleared everything off the table and mapped out the whole city, starting with the borders of the neighbourhoods, then the building plots followed by the streets. There it was: the new plan for a new city rebuilt from the rubble. He looked at it with pride, and then he started building, up and out. As Mohammed built he kept a handwritten note taped to the wall above his table. He read it often. 'They destroy. We rebuild.'

AFTERWORD

In the winter of 2015, as the war continued to get increasingly violent, Shah'd and Wael had their fifth child, Amir. At the same time, I made contact with the Qutaish family and began working with a string of volunteers across Syria, Turkey and America to safely transport a portion of Mohammed's model from his home in Aleppo to Mmuseumm in New York City to begin its tour as a public exhibition.

Shortly afterwards, the rest of Mohammed's model was damaged when a bomb hit the building next to theirs.

The world has watched as the war continues. Today it is estimated that over

400,000 people have died and 11 million, half the country's population, are displaced. It is considered to be one of the worst humanitarian crises since the Second World War.

In early 2016, Mohammed and his family fled Syria, leaving everything behind in pursuit of safety. They are currently living in a small town in Turkey, looking for ways to reach Europe or Canada safely and legally.

Mohammed still hopes to attend university in a few years time to become an architect, and to one day contribute to rebuilding his beloved Aleppo.

While his city was reduced to rubble by civil war, Mohammed dreamt of becoming an architect.

Public

Modern aviation began when the Wright brothers invented the airplane at Kitty Hawk in 1903. The statement is as plain and simple as a schoolroom fact. What is obscured in this succinct telling, and what is obscured in most people's minds, is the science and patience the invention entailed. 'Invention' sounds like a flash from heaven. But you do not wake up one day and make an airplane out of a mere notion. To create a heavier-than-air, human-piloted machine – to create such a machine so that it can reliably take off, undertake controlled flight and land safely – is to have solved a large number of difficult problems. In the unseen background was the brothers' boyhood fascination with birds and flight, their careful selection of the Outer Banks of North Carolina (they were from Ohio) for optimal wind conditions in which to trial gliders and the scores of different wing shapes they tested in 1901 and 1902. Invention is not easy.

The innovation behind Aquila – should it eventually succeed – is nothing as momentous as that of the Wrights. It will alter the world but not its fundamentals. But it is true that Facebook, under the aegis of its Connectivity Lab, is trying to usher into existence something that requires great feats of engineering, design and human problem-solving endurance, something that requires the bringing together of a number of things that have not been yoked into a single concept before.

Aquila is massive: it has a wingspan of 42.4 m, which is comparable to a Boeing 737. But because of the materials from which it is made – a great deal of it is high tensile carbon fibre – it is also surprisingly light at less than 453 kg, substantially less than the weight of a small car. It is an unmanned plane, to be piloted remotely. In other words, it is a drone.

That is the plane. The plan behind the plane is a goal of delivering internet connectivity to remote and underserved areas. What does a drone have to do with the internet? Facebook's hope is that the Aquila planes (the Aquilae?), powered by solar energy, will loop around in the air above a given territory for months at a time. They will fly at an altitude greater than commercial flights, above 60,000 ft (18,288 m), and will beam down invisible infrared laser beams that flicker on and off billions of times per second sending data at fibre-optic speeds to a transponder that converts them to a Wi-Fi or 4G network.

If this sounds futuristic, that is because it is. Little of it has actually been done before and none of it at this scale. No commercial venture has flown at such altitudes – only a few military planes do. The extremely low air density at such heights is a technical challenge. The longest time a solar-powered drone has stayed aloft, until quite recently, is two weeks: how will the leap to Facebook's goal of three months happen? Furthermore, it is difficult for lasers to pass through clouds. And this is all without even taking into account the other variables of flying drones in remote, impoverished or disputed territory, the geographical terrain in which there is little or poor internet coverage. There are questions of weather and of war. There are also political questions: in a techno-optimistic world, where every village priest, maid or fisherman has a mobile phone or two, why are there places still off the connectivity map? Can the problem be reduced simply to a lack of technical solutions?

Facebook's fleet of pilotless aeroplanes will stay aloft for months, beaming down
the internet to areas with poor connectivity.

Public

(Above)
Orville Wright flies his Type A aircraft at
Tempelhof airfield during the September
1909 Berlin Airshow.

(Opposite)
The 1903 Wright Flyer in the Kensington
Science Museum in London, 1928.

Public

Aquila is Latin for 'eagle' and the plan the Connectivity Lab offers is lofty in more senses than one. Facebook's CEO Mark Zuckerberg speaks of the project in the by-now familiar visionary cadences of philanthrocapitalism: 'Connecting the world is one of the fundamental challenges of our generation. More than four billion people don't have a voice online.'[1] The great good that this connectivity is meant to achieve is what we are asked to focus on. The prospect of improving human life by the billions is a dizzying one, and Zuckerberg already has the gleam of such achievement in his eye. But when the sky is full of the Connectivity Lab's envisioned 10,000 high-soaring aircraft, will it be possible to ignore the dark sides of this great technological leap? In an age of surveillance, data gathering, despotic governments and the persecution of dissidents, few things can be more ominous than a sky full of drones.

Given the recent debacle in which Facebook attempted to supply a free, but severely delimited, internet package through a project named Free Basics to impoverished areas in India, there is also the pressing question of who will handle the issues of local permissions and the rights of the beneficiaries The problem here is a perennial one: not everyone who is in perceived need wants to be saved on the saviour's terms. People are particular and some people would rather have no internet access than one dictated to them by an outside party.

Aquila, assuming it proves viable, will bring predictably good effects as well as unpredictably bad ones. What it cannot be is neutral. Unlike the Wright brothers' invention, which evolved into its current complexities over decades, this one comes heavy with the baggage of a company already worth billions of dollars, a company with enormous clout in the political sphere. What would a world according to the logic of Aquila look like? To a certain extent, that aquiline world is already here. There is a great deal more at stake than a single company's decisions, powerful as that company itself might be. The systematized dissemination of information (such as Aquila promises from on high) is difficult to separate in any meaningful sense from systematized data collection. It is always two-way traffic, and we already live in a world in which surveillance is common. Certain technologies, not explicitly designed for surveillance, are likely to help make it even more pervasive.

Guy Debord cites a certain map in his *Theory of the Dérive* (1956).[2] It is a map drawn by the sociologist Paul-Henry Chombart de Lauwe of 'all the movements made in the space of one year by a student living in the 16th Arrondissement'. This one woman, over the course of the year, has a limited trajectory, most of it a triangle between her university, her home and her piano teacher. What was a conceptually peculiar map about the movements of one student in the 1950s is now, for all of us, an inescapable reality. The complicated polygons of our movements over time, all over our cities and all over the globe, are already tracked by our mobile phones. As forms of visual tracking become more sophisticated (and they are already considerable, if we take into account security

Public

cameras in our cities and towns), and as data storage becomes ever cheaper, 'activity-based intelligence' will be the governing principle not only of the military sphere, where it emerged, but also the commercial and public spheres.

The coming age represents a collapse between the functions, goals and relative capabilities of the military–industrial complex and the civilian technological solutionists. If Facebook, Google and Apple (or other companies with similar reach and power) already have within their control more data about citizens than do most sovereign states, then they will naturally assume some of the functions of the state. Tech companies have, or will soon have, the power to withhold essential services and even the power to influence political processes. Facebook is no longer just a website for meeting friends. Google is far more than a search engine. They are building things, controlling actual geographies and becoming infrastructural facts on the ground and in the air.

One of the key philosophical battles of this already-future would be how to educate a gradually disempowered populace about the algorithmically driven technologies under which they now live. These technologies are not neutral, not only because of the biases of their programmers, but also because, at the level of complexity at which they work (billions of lines of code), they are totalizing systems. The new technologies, whether they are drones or wearable devices, quantify the information they produce, disseminate the information they quantify, and monetize and politicize the information they disseminate. They create a complex loop of data-based activities. Are we humans then using tech? Or are we simply terms in its symbolic logic?

Invention is the lifeblood of society and projects such as Aquila are impressive, but it would not do to greet their arrival only with applause. Questions must be asked that go well beyond optimism. What will citizen power look like in such a future – will it have any kind of independence from the whims of for-profit corporations? What will commercially driven technology mean in a time of climate change, water shortages and mass migration? Who will have control over the electrical lines, the sewage systems, the roads and the communications networks? When tech is king, what will constitute treasonous activity? What will be the fate of dissidents? The coming innovations cannot be taken at their word that they are merely humanitarian tech. They must be understood as both features and drivers of our

utterly transformed, and potentially dystopian, political landscape. It is good to remember that within a decade of its invention, the heavier-than-air airplane was being used to drop bombs on foreign enemies; and it of course went on to have an outsized role in the carnage of both world wars. The wonderful innovation, more often than not, comes with a sting in its tail, a sting that, even if not avoidable, can be anticipated.

(Opposite)
Mark Zuckerberg holding a propeller pod for Aquila at a conference in San Francisco, 2016.

Public

On 16 May 2011, Frank La Rue, the United Nations Special Rapporteur on the Promotion and Protection of the Right to Freedom of Opinion and Expression, released an investigative report recognizing the central and growing importance of the internet in people's daily lives and declared that internet access is a human right.[1] The report, published just months after the Egyptian government cut off internet access in the wake of growing protests in the country, noted that access has two key dimensions: firstly, unencumbered access to online content, and secondly, the availability of infrastructure that makes that access possible. For those living in parts of the world with ubiquitous, stable connectivity, it can be easy to overlook the sheer importance of this latter dimension.

With metaphors like 'the cloud' – a Silicon Valley buzzword for remote data storage – and access increasingly taking the form of Wi-Fi and mobile broadband, the internet can indeed seem like data floating through the sky. However, connectivity is built on very physical objects: mobile towers dot our urban and rural landscapes, humming server farms span hectares of former farmland, undersea and underground cables criss-cross our oceans and continents, and massive routing buildings coordinate connection points.

A better analogy to clouds might be potable water. Water, a ubiquitous and necessary good for human life, relies on vast pipes, sewage systems and manual methods to ferry safe and clean access to all. In a city like London, this means that almost every tap, when turned on, spouts clean water. In a rural village in Kenya, people might hike dozens of kilometres each day to pick up water from a

central water pump and ferry it back to their homes in large jugs. Internet access is not dissimilar. In urban Europe, the infrastructure for the internet, like water, can be invisible and taken for granted, turned on and off with ease. In rural East Africa, the infrastructure for the internet may be more visible in its early stages: mobile broadband towers emerge, electric wire is laid down and connectivity arrives. When that physical infrastructure is lacking, people find other ways to ferry digital content such as movies and music, which are then carried back home on their phones or USB sticks.

Enter BRCK, a rugged mobile Wi-Fi router designed and produced in 2013 by the eponymous Nairobi-based company. In contrast to 'the cloud', its name more directly evokes the internet's physicality and is geared for use in the global South. A 72 by 132 mm black box that is 45 mm thick, the BRCK sells for $250 and looks and handles like its namesake. Sturdy, dustproof, water resistant and surge protected, it can handle challenging natural environments and the failures of human infrastructure, like dirty voltage connections. BRCK takes a basic 3G SIM card and broadcasts that mobile connection to up to 20 devices. Barring access to a SIM, it can even connect to a virtual global mobile network. The device features an extendable antenna to maximize access to wireless networks, a variety of ports for direct connections to a solar panel or car charger, and even 4GB of data storage.

This latter piece – a hard drive – is an innovative feature. The water metaphor helps to illustrate its importance: imagine a context where internet connectivity, like water, spurts and sputters inconsistently during the day, when everyone is

The BRCK ('brick') is a mobile Wi-Fi
router and hard drive that provides
access to the internet in zones with
reduced connectivity.

Public

using it, and flows a little more smoothly at night, when people are asleep. This reflects the experience for many parts of the world with limited internet infrastructure, especially in the global South. If people could fill up a bucket of data when internet connectivity is stronger, they would have plenty to use and reference during the day without the frustrations of spotty connectivity.

BRCK's hard drive functions like a bucket of data, accessible whether or not the BRCK is connected to mobile internet. As mobile connectivity can be unreliable and sometimes non-existent in rural areas, the built-in hard drive is a simple but transformative feature. In 2015, the company developed the Kio Tablet, a rugged tablet designed explicitly for developing regions, which can siphon data directly from the BRCK. When connectivity is strong, the BRCK hard drive can be loaded up with educational materials – books, videos and audio files – that can be shared to tablets within a Wi-Fi intranet. This transforms the BRCK into a central source of data later, even when a connection to mobile broadband is not possible. At night, when bandwidth is stronger, the BRCK can once again download new materials. This store-and-release approach to data is quite common in developing regions of the world with intermittent connectivity, and BRCK and its programmable interface makes the process more seamless.

IN THE FIELD AND IN CONTEXT

'We'd be out in the field for basically the entire day', noted design researcher Zachary Hyman in an interview about a recent research trip he took to the Jordanian desert. Hyman travels frequently in developing regions to conduct research, and he and the team he worked with relied on the BRCK to provide regular communications with their United States office. 'We were on the move for about 16 hours. There wasn't always time to slow down and plug in the BRCK so it could be charged.'[2] The device's durability helped them weather the dusty environment and even a sandstorm. Furthermore, it allowed them to charge their phones via its USB ports and to quickly and easily update and sync design prototypes in the field for rapid feedback.

Back in Amman, the BRCK proved its versatility as a Wi-Fi extender rather than a mobile router, helping the team access their apartment's internet in spotty corners. This begs a question: which Wi-Fi were they extending? In London, Wi-Fi might

BRCK on the go. Paired with a signal booster and an antenna, it reaches wireless internet.

Public

emanate from a standing, blinking router plugged into a wall socket connected to an underground fibre-optic cable network. In Amman, mobile broadband is a much more common access point. The apartment was equipped with a Huawei mobile broadband modem, which did much the same work that the BRCK does: transforming a mobile broadband signal into a usable Wi-Fi signal for multiple devices.

Produced by companies like Huawei, ZTE and Netgear, mobile broadband modems sell in the millions each year, both as stand-alone devices and bundled with service plans. Starting at US$30 and often designed to fit in a small bag or pocket, their price point is much more accessible to people with limited incomes. Their discreet size makes them conducive to everyday access without drawing too much attention. Indeed, these devices arguably have become as central to internet connectivity infrastructure in low-income parts of the world as wired routers have in high-income areas. Like the BRCK, many of these modems can connect to external hard drives designed for Wi-Fi intranets and to mini solar panels.

Around the world, mobile broadband is filling in for wired infrastructure, driven by a growing global demand for connectivity. The world overall is moving towards a 'connected majority', wherein most of the population will have regular, individual access to broadband internet. We will be crossing that threshold in a matter of years: a 2015 report by the International Telecommunications Union noted that some 44 per cent of the world's population are now internet users, up over 800 per cent from the year 2000, and the majority are on mobile.[3] Those who access the internet via shared accounts and devices are not necessarily covered in this report, and so the reach of the internet may be considerably greater.

In this context, BRCK serves as a specialized device in at least two ways. Firstly, its consolidation of key technologies in a durable product makes it extremely reliable for fieldwork and research. The combined hard drive, router, antenna and surge protector reduce the number of complex integrations a mobile team needs to worry about. Secondly, birthed out of Ushahidi, the Nairobi-based social enterprise made famous by its mapping software used in disaster recovery efforts, BRCK the company has larger educational and advocacy aims. Its work to provide a holistic educational solution with the Kio tablets has the potential to transform rural educational initiatives by simplifying the process for educators and ameliorating the need for costly technical training and setup time.

Although the device is clearly positioned for use in the developing world, BRCK has also received backing from funders in wealthier countries, such as former AOL founder Steve Case and Synergy Ventures.[4] BRCK's founders speak in western venues like TEDGlobal and the Massachusetts Institute of Technology to illustrate the context in which the BRCK is used and the potential impact of the company's work. Growing awareness of BRCK and similar devices has the potential to expand the conversation in western media, and in academic and policy discourse, about what connectivity can look like under diverse conditions.

If we are to make internet access a truly universal human right, this awareness-building is critical. How the next half of the world connects to the internet will look very different from how the first half has connected. This next half will be more mobile and more rural and have extremely different levels of literacy, income and living standards. Designers and technologists will need to think more creatively about the very infrastructure of the internet and reduce the road bumps to connectivity, whether through mobile, wired or as yet unexplored technologies. BRCK may be just the beginning.

THE CAMPAIGN TO STOP KILLER ROBOTS

Richard Moyes

The Campaign to Stop Killer Robots is working to shape the future of weapons design: to prevent the development of systems in which computers and sensors can independently 'decide' where violent force should be applied. The Campaign is an international coalition of non-governmental organizations that includes roboticists and ethicists, as well as disarmament, human-rights and peace activists. It was formed in New York in October 2012 by a group of 10 organizations and in the context of a growing sense of urgency in civil society to challenge increasing autonomy in weapons systems.[1]

Rapid advances in computing, artificial intelligence (AI) and sensors are bringing the possibility of 'autonomous weapons' closer all the time. How the international community should respond to these developments, and potentially challenge and restrict such technology in the context of the use of force, is under discussion at the United Nations. This is taking place within the framework of a convention mandated to apply restrictions and prohibitions to weapons. At the time of writing, the future of that work is hard to predict. Yet they are discussions that draw out critical themes regarding the relationship between technology and society.

There are numerous issues in this debate, but three important themes are: how the design of future autonomous weapons systems may serve to embed and encode categorizations of people and things in society; how certain arguments made in favour of autonomous weapons point to a more disturbing vision of the 'perfection' of violence; and how the concept of what is 'inevitable' is used to argue around the prospects for us shaping the future of technology in the way that we want.

CONTROLLING WEAPONS

The track record of international efforts aimed at limiting and controlling technologies of violence has been problematic. Over the last 100 years, broad general rules have been developed to limit the use of violence in armed conflict, such as the Geneva Conventions that constitute a significant part of the restrictions on conduct in war. Among the so-called 'weapons of mass destruction', chemical and biological weapons are prohibited under specific treaties. Only in July 2017 was a treaty prohibiting nuclear weapons adopted at the United Nations; it will take some time for that treaty to enter into force and nuclear-armed states currently reject it. Where other (so-called conventional) weapons have been outlawed, this has often happened only after significant levels of civilian harm have already been incurred (as was the case with anti-personnel landmines and cluster munitions). Establishing a ban on something before it comes into existence presents distinct challenges: of building motivation, of obtaining agreement on the problem and of shaping a legislative approach to an uncertain future.

IMAGINING KILLER ROBOTS

Autonomous weapons, or 'killer robots', are imagined in a variety of different ways. Images of the Terminator from James Cameron's 1984 film and subsequent sequels tend to dominate media

Public

How do we shape a legislative approach
to an uncertain future? Mary Wareham
and the Campaign to Stop Killer Robots
participate in UN talks about autonomous
weapons.

Public

The Pentagon has plans to expand
combat air patrol flights by remotely
piloted aircraft, like this MQ-9 Reaper,
by as much as 50 per cent.

Public

representations of the issue, pulling people towards the idea of humanoid human replacement: self-contained, interacting on human terms and roaming unassisted in pursuit of a goal. Such futuristic creations may capture our attention, but they distract us from the present. Weapons systems are already deployed that use sensors to assign categories to the outside world. Through algorithms, they make decisions about which signals match what has been programmed as a target and so they determine where force will be directed. From such systems we have the conceptual building blocks of real killer robots: systems – whether in the air, on land or at sea – that employ sensors and computing independently to identify and apply force to targets in the world. But how such targets are encoded, and what such a process means for our understanding of our place in society, can present a dark vision of the future.

ENCODING

A characteristic of autonomous weapons, of whatever level of complexity, is that they must rely on certain categorizations of people or objects in order to assign those things, ultimately, as targets or something else. Such a process inevitably requires formalized representations of people or things to be analysed in the stead of the thing itself. These proxy indicators are pre-programmed descriptions of what is to be targeted, boiled down to terms that the weapon's sensors can detect. So, for certain weapons systems already in existence, a heat source of a certain shape becomes an indicator of a vehicle of a certain size that becomes an indicator of an enemy tank, which is considered to be a target. Developments in sensors, biometrics, computer power and algorithms all open up diverse possibilities for how future weapons might use different representations to encode targets. For people using simpler weapons in combat now, whether shooting a gun or firing a missile from a drone, similar processes of instinctive categorization are taking place – yet the person going through that process is doing so aware that they are observing a real vehicle or a real person, and that they are making a moral choice in their decision to fire. The proxy indicators for autonomous weapons are not acted upon in a situation of moral doubt. They simply represent labels in a computerized bureaucracy: one that will apply force facelessly based on fixed categorizations established in advance.

THE PERFECTION OF VIOLENCE

This bureaucratization of violence has stark implications. There are people who argue that handing over such life and death decisions to machines may be preferable to leaving them in the hands of humans: humans make mistakes, they are prone to anger and might act selfishly or out of fear, they may even commit atrocities. Computers, however, will not suffer these human failings. Such perspectives suggest that we might be able to design out the human weaknesses, to be left with something mechanized, but surgical and clean.

Such a vision is often countered by arguments about the likely limitations of technology: machines will never have the processing power and sensitivity that proponents claim. Yet speculation about future technological capability, and discussions about likely outcomes, tend to mask more subtle problems in such arguments for the superiority of autonomous weapons: that they suggest a perfectibility of violence when violence is itself a human failing.

The idea of designing human fallibility out of the enactment of violence is simply another way of encoding and then asserting the overarching power of a particular human bureaucratic structure, with its own biases and fallibilities.

While individual humans are known to have committed atrocities in conflict, it has been the bureaucratization of violence, the labelling of people into set categories for systematic processing, that has brought forth the worst of human society: in acts of discrimination, persecution and genocide. Although those acts may seem alien to us now, they were enabled by shared understandings that certain labels could be applied to certain groups – by race, ethnicity or other social indicators – and that this was a reasonable basis for making them targets. Sufficient people had confidence in the validity of those categorizations, in their historical context, for bureaucratic processing of these groups to be accepted. We should be wary of claims that our own societies have developed better systems for categorizing people to be the recipients of violence – a tendency that autonomous weapons would likely encourage.

INEVITABILITY

In the face of developing technology, ownership of the future turns out to be a critical point of rhetorical engagement. Numerous commentators opposed to drawing a line against the evolution

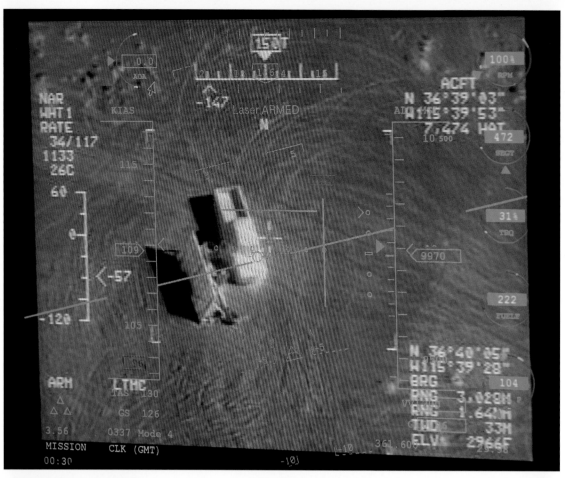

Public

of fully autonomous weapons argue from the position that their advent is inevitable: that technology is simply proceeding in this direction and that acting as if it could be avoided is not only futile but also risks the advantages that technologically advanced nations might be able to achieve in this area. There seems little doubt that the component technologies that would enable autonomous weapons will grow independently. Already in armed drones – remotely piloted vehicles from which weapons can be fired – we have a platform for the delivery of force under the control of an electronic signal, currently controlled by humans. Elsewhere we have sensors and algorithms capable of identifying and proposing certain categories of things as targets, which are then sent on to humans for subsequent action. Joining up such systems to remove the need for human intervention is not a technologically distant challenge; in engineering terms it is barely a challenge at all.

Yet to argue that technological possibility results in something being inevitable is just another device to erase human agency from our relationship with technology and with violence. The design and construction of weapons systems is still a human process, and humans have the capacity to deliberate on whether the structures of violence that they are enabling are appropriate to the society in which they want to live. They are also capable of shaping broad social conversations that help people to frame those personal choices: conversations that can find stronger form when they are embedded in our policy and legal frameworks. For the Campaign to Stop Killer Robots the assertion is perhaps quite simple: we should ask our governments to reject this future of violence and to ensure that weapons will remain under meaningful human control.

(Opposite above)
A truck as seen from the eyes of an MQ-9 Reaper, captured during an August 2007 training mission at Creech Air Force Base in Nevada.

(Opposite below)
A ground-control station cockpit used to control remotely piloted aircraft, Creech Air Force Base, November 2015.

Public

PLANET

We live in a time of deep ecological mutation, which scientists now define as the 'Anthropocene'. This expression emerged in the 2000s to describe a new geological epoch marked by human activity and its effect on the planet. Having shaped the planet to a point where our own species might not survive in the future, should we once more intervene in the earth's system in an attempt to repair this damage?

Aerocene Explorer
Tomás Saraceno
2016

Inflated by air, lifted by the sun and carried only by wind, this balloon can be used to collect atmospheric data and take aerial photographs and videos. Part of a larger research initiative, Aerocene Explorer hopes to help achieve a more sustainable future where people will live in the skies on floating cities.

Planet

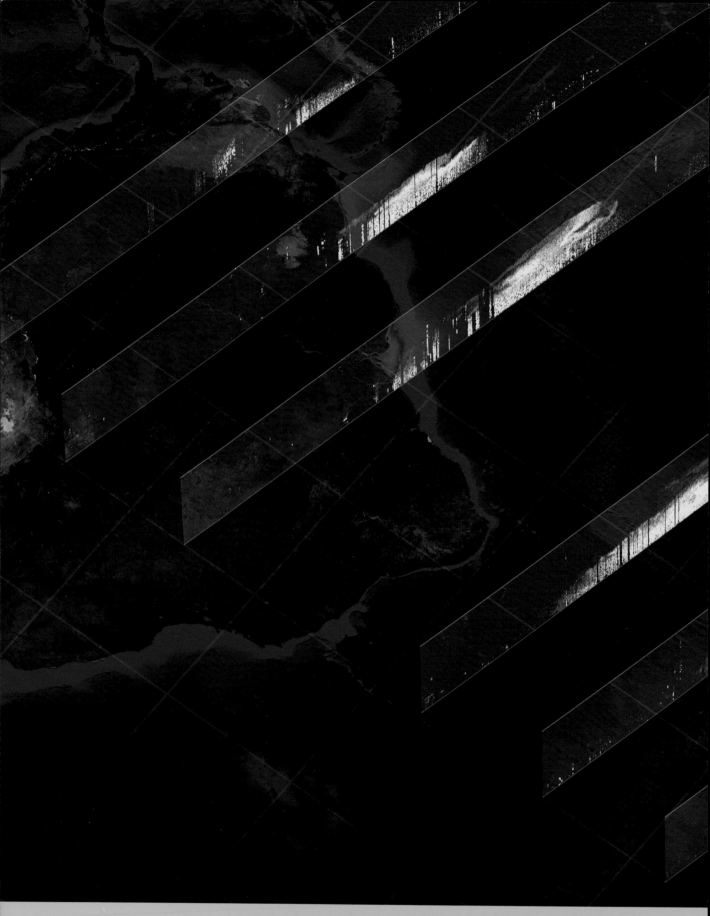

Saharan Dust Crossing the Atlantic Ocean
NASA Scientific Visualization Studio
2010

This image reveals how much dust makes the trans-Atlantic journey from the Sahara
Desert to the Amazon rainforest. Among this dust is phosphorus, an essential fertilizer,
on which the Amazon depends in order to flourish. This finding is part of a larger research
effort to measure the circulation of dust and aerosols, and their impact in global climate.

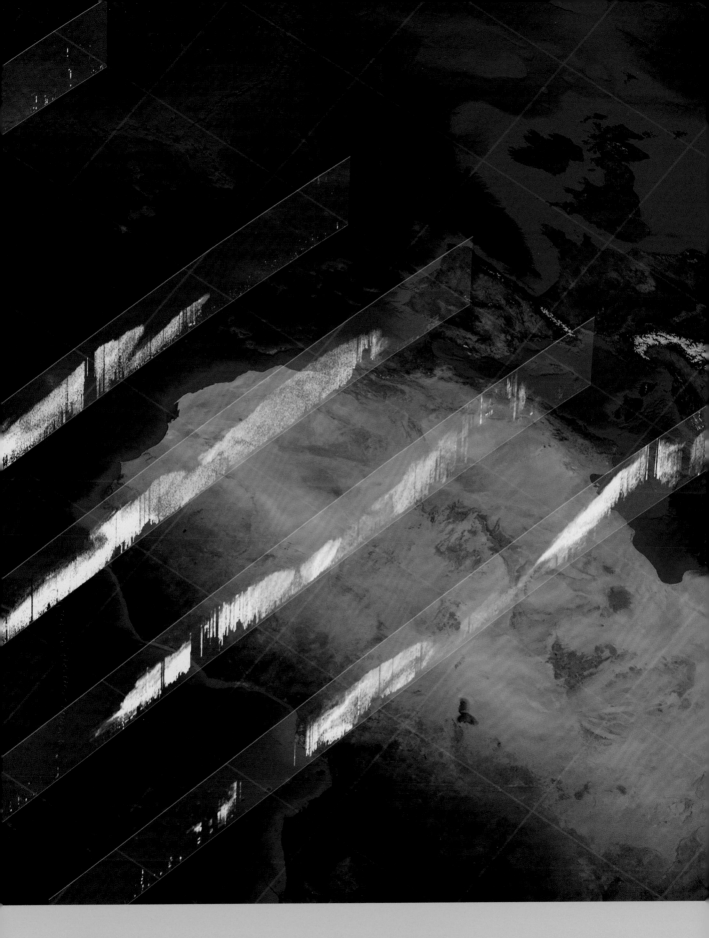

Planet

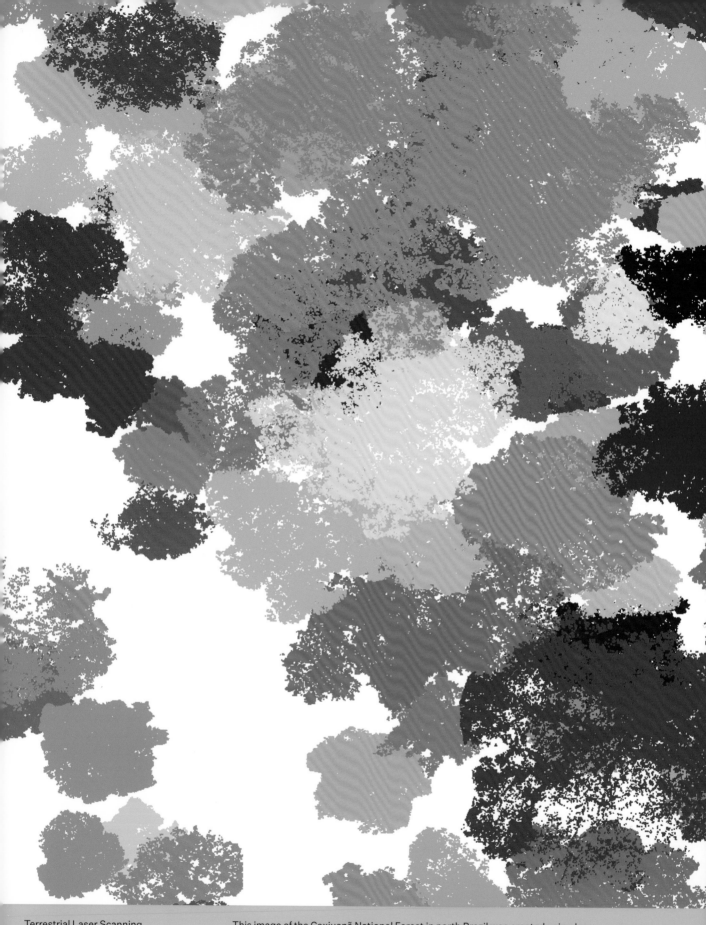

Terrestrial Laser Scanning
Dr Mathias Disney,
University College London
2017

This image of the Caxiuanã National Forest in north Brazil was created using laser scanners measuring the precise size of trees. The information gathered is then employed to improve the accuracy of evaluations of the amount of carbon biomass stored in forests, data that could potentially be used to halt deforestation.

95   Planet

CubeSat
Clyde Space
2015

Small and cheap, CubeSats are mini satellites that allow private companies, academic institutions and even individuals to conduct research in space for a fraction of the cost of a full satellite. CubeSats have been used to detect earthquakes, study climate change and space weather, and search for planets in other solar systems.

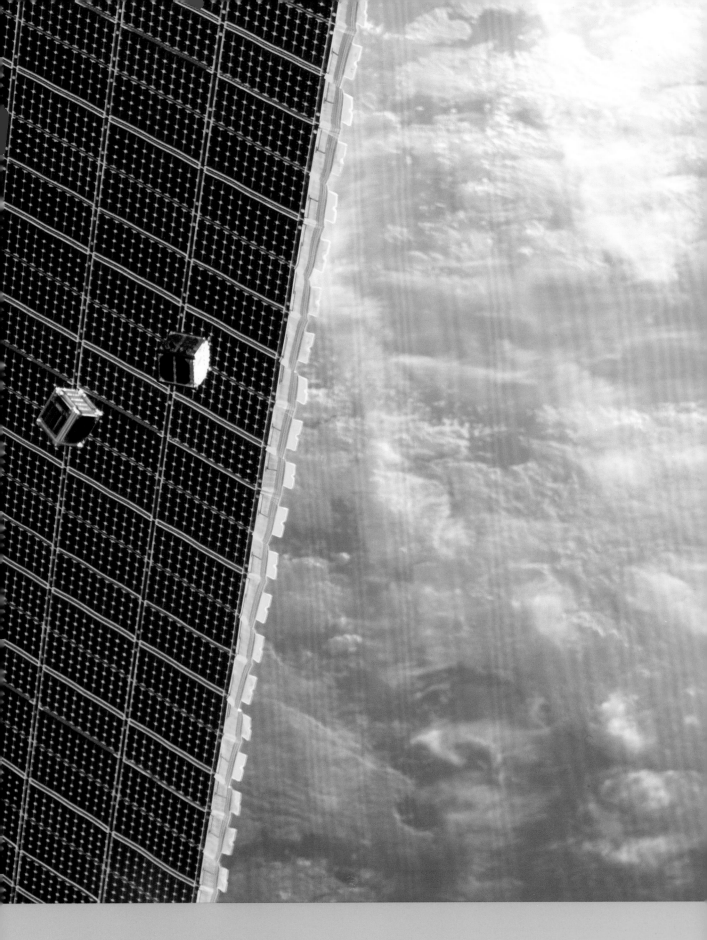

Planet

Falcon 9
SpaceX
2011

A rocket lands successfully back on earth having delivered a satellite into orbit.
This reusable technology has radically lowered the cost of reaching space. For
Elon Musk, the founder of SpaceX, this is a step towards an interplanetary future.

Planet

'This is not order
but the absence
of order.

He was wrong, the unanswering
forest implied:

It was
an ordered absence'[1]

These lines, taken from Margaret Atwood's 1978 poem 'Progressive Insanities of a Pioneer', tell of the colonial protagonist's attempts to understand and position himself against the unending forest before him; 'on a sheet of green paper / proclaiming himself the centre'.[2] Eventually, the forest speaks back, implying that it is not the absence of order as the pioneer originally read it, but something more complex, a system beyond his perception. The vacant space between the trees was not silence as the pioneer knew it to be, but instead a more complex system he could not hear, or see.

In 1997, Suzanne Simard, a professor of forest ecology at the University of British Columbia, revealed the secret conversations between trees and the fungi that live between their roots.[3] Once considered parasitic, such fungi are in fact a symbiotic ecology, mutually dependent on this process for their own survival. Each tree is connected to many others, bridged by the fungi which forms a complex communications network. One tree affects the lives of the others, even if they don't appear to be related at first sight. Simard's discovery was based on observing the removal of a fir tree, only to witness a seemingly untouched birch die with it.[4] Connections exist where they are not immediately obvious or acknowledged, and the disruption of one part of a system, one node, has the potential to have catastrophic impact upon others. If not now, then perhaps later, when it is harder to see where it broke in the first place. Nothing exists in a vacuum; the forest is a relational system of communication.

Jalila Essaïdi's Tree Antenna speaks to that which we cannot see. It taps into the existing system of the tree, transforming it into a communications network that humans can use. By wrapping a coil of insulated wire around a tree trunk, hooked up to a toroid conductor, it enables the broadcast and reception of longwave radio signals across seas and territories, amplified by the tree. Alone, it is an antenna, while many together become a living communications network. This network is a parallel system, independent of the one network we have come to depend on today, the internet. In this way, the project quietly propels us towards a future in which covert, undercover communications are necessary or preferred.

Such a system would become necessary following the potential collapse or constriction of the systems running the internet as we currently know it. Already a hugely contested space, the internet sometimes feels more like a battleground than the paradise we were promised, with the presence of corporate chokeholds and government restrictions a consistent backdrop to daily use. Not everyone uses the internet in the same way, and many still benefit from its positive and affirmative effects, yet all encounter the signals of something darker lying just beneath the surface. The internet, once a Californian dream

By wrapping a coil of insulated wire around
a tree trunk, Jalila Essaïdi's Tree Antenna
enables the broadcast and reception of
longwave radio signals.

Planet

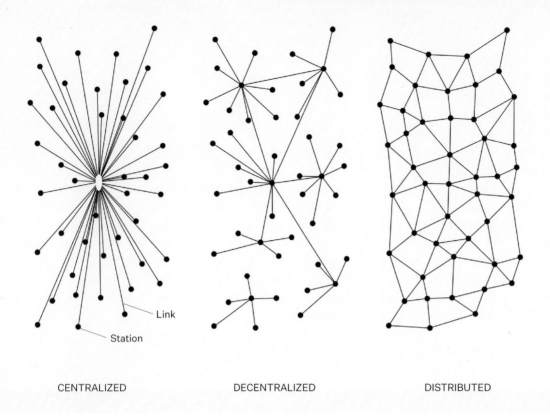

Link

Station

CENTRALIZED   DECENTRALIZED   DISTRIBUTED

of freedom, has become an arm of control, an incendiary tool of capitalism. The repeal of net neutrality in the US is, hopefully, an early warning sign to those who only saw the internet as one fluid pipeline.

To imagine the end of the internet, or the end of an internet in which we are able to operate with some freedom, forces us to consider the extent of possible consequences of such an event. It forces us to admit our dependency on the internet and consider alternatives to it. We only have to look to China to speculate on what this might be like, where extreme censorship and state run services already restrict, monitor and control users. Some may accuse these speculations of being vaguely post-apocalyptic, and perhaps they are, but sometimes it is important to imagine the end of the world, or at least what we know the world to be. It is maybe useful to rehearse the final days, the last hour. Ctrl + Alt + Delete. Where could we communicate outside of these closed spaces?

Jalila Essaïdi's project is, in many ways, a preparation for an ending to the internet; creating points of resilience and resistance in which to break, fast-forwarding to a future where communications are fundamentally changed. I talk from a Western,

privileged, perspective here, though we cannot ignore the fact that it's not just direct internet users, and those able to access it, who are affected by its existence. Those submarine cables have to run through somewhere, the energy churned by data centres affects someone's coastline, someone's sky.

Essaïdi's Tree Antenna offers a possible alternative to current modes of global communication, and a potential return to the grassroots, community led ideals that first thrived on the internet. It departs from the findings of Major George O. Squire of the US Army Signal Corps. As he wrote in a 1919 report to the Department of War,

> It would seem that living vegetation may play a more important part in electrical phenomena than has been generally supposed. If as indicated in these experiments, the earth surface is already generously provided with efficient antennas which we have but to utilize for communications.[5]

The report, published in the *Journal of The Franklin Institute*, revealed the ingenuity of these communications pioneers. The landscape had instantly transformed, the forest had become a tool beyond a means for cover and camouflage.

In 1969, the US Army Electronics Command borrowed this discovery from Squire and his team, using trees in New Jersey to test short and medium wavelength radio communication. At the end of the 1970s, the same unit conducted the Panama Canal Zone Experiments, with a 'Performance of Trees as Radio Antenna in Tropical Jungle Forests',[6] a research and development programme in a location far more unknown, and far wilder. Their toroid conductor inside a thick plastic insulation was, like Essaïdi's, wound into a coil, using the tree as a Hybrid Electromagnetic Antenna Coupler (HEMAC), previously found to enhance high and medium frequency signal emissions wherever they needed to go.

Despite the efforts and ingenuity of these experiments, the change of environment proved too much. The rainforest was too wet, too soft, and too rotten for clear communication, and far too complicated to manage. The signal could be heard, but was not to be relied upon, recommended for use as a backup only. Instead, they proposed the antenna could be used to control 'remote and submerged devices';[7] in less abstract terms, the activation and deactivation of land mines. In attempting to harness the rainforest, the researchers deferred to a more direct form of weaponization. In the appendices of the report, several men are seen grinning against lush vegetation, clinging to bark that crumbles beneath their apparatus.

Now, almost one hundred years after Major Squire's letter, Essaïdi once again returns to the vision of forests as fields of antennas for communication. With this new infrastructure comes a new way to talk with one another, to send information that could be silenced in a future where the internet is no longer an option for free communication. But unlike her predecessors, Essaïdi does not have military aspirations for her experiments. Instead, she is an artist and entrepreneur exploring the intersection between art and biology. Famous for her 'bulletproof skin' project that combined spider silk proteins and human skin cells, Essaïdi's key design contribution is biological.[8] Her tree antenna has one subtle difference to the experiments of the twentieth century. Where Squire, and the teams that followed, used nails to hammer into the trunk, Essaïdi's model does not harm the tree. This particular, thoughtful, consideration is a result of looking at all that could be at stake. Leaving the tree untouched (while still gaining significant results), Essaïdi acknowledges the existence of an inherent system to the forest.

Although it is not quite right to compare the internet to a forest, the analogy can help to understand the nature of complex systems. Both are vast, both have levels beyond what the naked eye can see, and both can, and perhaps inevitably will, break. Could the same mistakes occur with Essaïdi's alternative that exist within the internet today? Could similar controls and restrictions arise?

Like Atwood's unnamed pioneer, it is not enough to read a new landscape in our own language, with our own names for things and as we understand it from our perspective. We must start to comprehend the world in new ways. This shift occurs by acknowledging the existence of a system beyond what lies immediately in front of us, and knowing what holds it in balance, what causes it to fracture, and what other things it depends on to exist. That which we do not see, and often, cannot know. As humans we are used to somewhat blindly navigating our place in a network of different, clashing systems, encountering one thing at a time. Perhaps, even if we cannot identify the beginning and the end of a given system, we can at least perceive the magnitude, and start to look below the surface, to all that belongs.

(Opposite)
Illustration of network types from Paul Baran's 1964 paper 'On Distributed Communication Networks'.

Planet

The radical nature of nuclear materials has consistently presented 'design' with its ultimate challenge: the problem of creating secure and stable storage for radioactive waste. This is a task that is complicated by the presumption that we live in a precarious and unpredictable world in which external dangers of one kind or another will most certainly unfold at some point in the future. In contrast, the design and engineering of nuclear reactors are organized by the belief that one can minimize risks, whether brought about by natural disaster or the result of human error, through careful planning and preparation. While reactors are not expected to operate beyond 60 years, the extreme temporality of their by-products – spent nuclear fuel – requires a design strategy that can guarantee containment of hazardous waste into the far distant future, that is, for up to 10,000 years, the date at which most anthropogenic radioactive isotopes will have fully decayed and discharged their lethal latency. Nuclear waste storage thus accepts the inevitability of change over millennia, whereas nuclear reactor design assumes the improbability of a highly unlikely event occurring during its working life.

Built around 2560–2540 BCE, the pyramid complex at Giza offers a useful design prototype for considering the problem of developing a long-term storage facility for preserving the radiological afterlives of nuclear waste into, what is effectively, eternity. Although clearly one of the oldest standing repositories for safeguarding precious materials – namely the pharaonic bodies of kings and queens – these ancient pyramids were frequently breached and their contents looted. Nor was the physical landscape into which they were built immune from transformations brought about by changing climatic conditions and human activities. While the original site of the pyramid complex was geographically removed from dense patterns of settlement, today it is located in the direct vicinity of upwards of 7 million people as the urban sprawl of Cairo encroaches. Such human immediacy and its attendant environmental impacts and industrial stresses naturally bear upon the material integrity and resilience of the pyramids despite their once robust engineering. Moreover, the hermetic and unadorned character of these monumental structures – devoted to preserving their pharaohs in a state of eternal repose – has not held vandals, fortune hunters, archaeologists, vendors and tourists at bay but, on the contrary, has drawn them all ever closer. As an ancient architecture of enclosure designed to stave off external threats, the variable history of the pyramids at Giza offers a cautionary tale for considering the challenges faced by contemporary efforts at locating, designing and managing geological repositories for the storage of intermediate-level nuclear waste. Few other human-made structures have ever attempted a similar feat, which is to say, to shield the material remains of the past against the volatility of unknown future events. That is, until ONKALO.

In 1997 the Joint Convention on the Safety of Spent Fuel Management and on the Safety of Radioactive Waste was adopted under the auspices of the International Atomic Energy Commission (IAEA):

> The Joint Convention is based on the
> general Safety Fundamentals for the

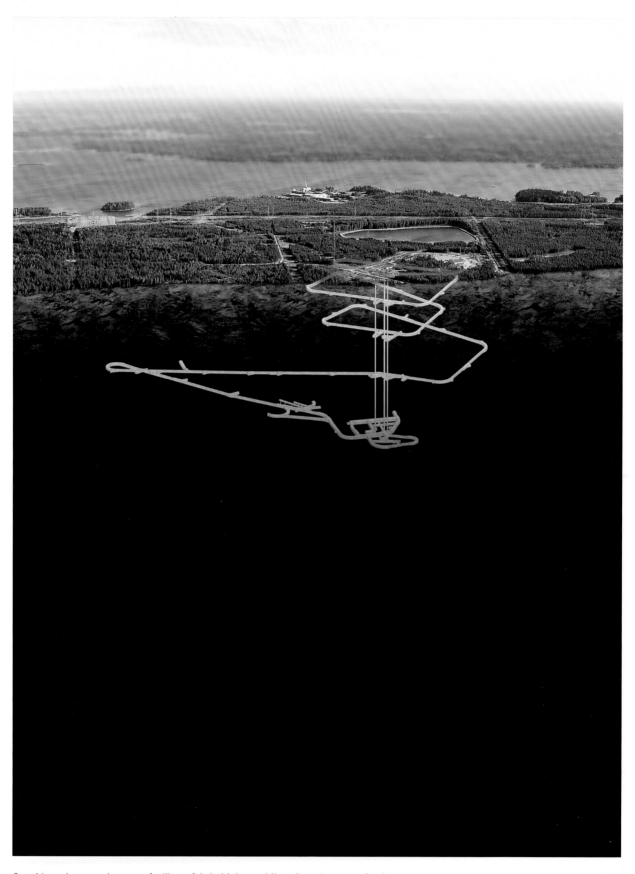

Can this underground storage facility safely hold the world's radioactive waste for the next 10,000 years?

Planet

implementation of nuclear waste management defined earlier by IAEA. These Fundamentals emphasise the responsibility of the producer of nuclear waste for implementing its nuclear waste management in a manner that allows humans and living nature to remain unharmed by the waste, in the present as well as the future.[1]

This Convention mandates that all EU member states are legally obligated to deal with their own nuclear legacy unless they have some prior arrangement in place. Having previously exported its spent fuel to Russia, Finland's Nuclear Energy Act was amended in 1994 resulting in the strict prohibition of the exportation or importation of nuclear waste. Consequently, a local solution was required for the disposal of radioactive materials produced by its Olkiluoto and Loviisa nuclear power plants. Directly adjacent to the Olkiluoto site, the world's first deep geological repository – ONKALO – is being constructed: a subterranean wormhole that will spiral 520 m down into the earth.

The aptly named repository, which means 'cavity' in Finnish, advances a design strategy that claims to offer the world's safest and most secure solution for the disposal of irradiated waste given the combined stability of the region's geology, groundwater conditions and remote island location. While Finland's political landscape certainly cannot be said to be unchanging, its natural history is considered immutable. Unlike the active seismic ecology of Japan, this area of Finland benefits from a steadfast geological character, deemed essential for sealing off nuclear hazards from external disturbances that might unleash their evil spirits in times to come. As with all nuclear waste disposal initiatives, unforeseen political events cannot easily be factored into the decision-making process of selecting a site or designing a repository, especially when multi-generational risk analysis is required. Assessments tend, instead, to focus on potential destruction or disruption from natural phenomena, such as earthquakes and tsunamis, rather than on externalities caused by war, industry or population expansion. Yet as the recent accident at the Fukushima-Daiichi Power Station forcefully recalls, efforts to control short- or long-term risk are ultimately a calculation that amortizes the contingencies of the future against the known perils of the past. Internal events – leaks, explosions and fires – are much easier to plan for than dangerous

events that might erupt out of unknown scenarios coming from the distant future.

The ONKALO facility has been conceptualized in four phases and began in earnest in 2004, although feasibility studies for the creation of a Finnish repository started in the 1980s. Screening processes were carried out from 1986 to 2000 and in the end two sites were selected: Eurajoki and Loviisa. Given that much of the material requiring disposal was already on the island of Olkiluoto, the decision to build a deep geological repository within the municipality of Eurajoki was determined to be the most economically viable as well as environmentally sound. Posiva, the same company that owns and operates the Olkiluoto Nuclear Power Plant, is the developer behind the ONKALO megastructure. With its bedrock research in hand, environmental impact assessments met and access tunnel now completed, construction is underway to build the three shafts out from which the maze of disposition tunnels will eventually branch. The scale of this sprawling subterranean architecture is impressive, boasting some 2,800 boreholes that will admit their radioactive bounty across 42 km of tunnelling distributed over a 3–4 km² region.

> In Posiva's reference solution, the canisters will be placed in holes 6 to 8 metres deep that will be bored in the floor of the deposition tunnels. The holes will be sealed with pre-compressed bentonite clay [absorbent clay formed by the breakdown of volcanic ash]. Alternatively, the canisters can be placed in horizontal tunnels, lined with a bentonite structure. During [the] final disposal operation, deposition tunnels are sealed as canisters are placed in [the] tunnels. After placing the canisters in [the] tunnels, the tunnels will be filled up as soon as possible. Compressed clay blocks will be used as filling material.[2]

ONKALO will go into full operation with the arrival of its first spent fuel canisters in early 2020. Prior to its final disposal, the spent nuclear fuel contained within must undergo a cool-down period of 40 to 60 years. This takes place in an intermediate off-site facility. From there the fuel is transported in purpose-built casks to ONKALO's above-ground encapsulation plant. Once the waste is transferred and packed into permanent storage canisters, these are inserted into the boreholes located deep within the bedrock. One hundred years later, in

'An inverted monument'. A section of
ONKALO's 52-km underground tunnel
network bored into granite bedrock.

The oldest Egyptian pyramids are only
4,600 years old. They were designed
to keep intruders out. Instead, they have
drawn people in.

Planet

approximately 2120, the disposal facility will be permanently sealed and will, ideally, withdraw from public consciousness, away from prying eyes and unwelcome intruders. As a megaproject and planetary-scale design object the sole ethos of ONKALO is rather *atypically* that of total and absolute disappearance: an inverted monument to a powerful nuclear remainder. At the same time, the lessons learnt from Giza stand as powerful reminders that mythic design objects organized around concealing their 'sacred' contents in a state of immortal suspension always compel rather than repel us. Rumours of an underground chamber of mineral wealth may well lure future bounty hunters or prompt new forms of resource extraction despite ONKALO's deliberately uninflected surface features. Ultimately, the soaring display and foreboding presence of the great pyramids did little to safeguard their contents. Will a counter-strategy of aesthetic innocuousness prove more successful? Only time will tell.

Comparing a geologic repository such as ONKALO with the funerary practices of ancient Egypt finds an analogy in the writings of Paul Virilio on the origins of the accident.[3] Virilio argues that technical accidents never just happen but are always invented as part of the creative design process: the train wreck invented with the design of the steam locomotive. When English archaeologist Howard Carter discovered the only intact burial chamber in the Valley of the Kings in 1922 – another geologic repository – and unearthed Tutankhamun's tomb 'he literally invented it', writes Virilio. Bringing something to light that was previously hidden is akin to inventing it anew.[4] Despite the actual existence of the tomb for millennia within the rocky basin of the Valley, the event of Tutankhamun's tomb can be said to have been invented in the twentieth century with its public disclosure. However, by way of contrast, when the reactor core at Chernobyl went supercritical, what was invented was not something entirely unknown but rather a technical latency or dormancy in the form of a 'major nuclear accident'. According to Virilio, Chernobyl was an accident lying-in-wait, one that was invented at the moment that Enrico Fermi first induced a nuclear chain reaction in the Chicago Pile-1 on 2 December 1942. Virilio forwards these two examples – the discovery of Tutankhamun's tomb and the accident at Chernobyl – to emphasize the degree to which the potential failure of technical objects is an inevitable by-product of their invention and use, and not merely the retroactive findings of some safety

procedure gone wrong or even the outcome of a natural disaster. While one can argue that Carter created the phenomena of King Tut and the Egyptomania that ensued in unearthing the tomb, one cannot say that the event of Chernobyl was simply invented on 26 April 1986 when its operating staff made a series of grave mistakes. Many prior factors contributed to this accident, not the least of which was the invention of the technology itself and the risks that attend any use of atomic energy as well as those associated with the disposal of its spent nuclear fuel. There are no 'accidental catastrophes' of a technical nature contends Virilio, which retroactively reveal a manufacturing defect, programming glitch or human error that led to the improbable event. Failure is always already engineered into the design process as technology's virtual double – its evil twin – the accident invented simultaneously with the invention.[5] This is why fail-safe procedures, emergency systems and contingency plans are required. The unthinkable future-event may not necessarily arrive in our lifetime, but given the extended temporality of nuclear materials, 10,000 years of flawless operations for a waste repository is still a considerable gamble.

At present the technical odds favour ONKALO when compared with other attempts at securing a future free from the threat of an accidental release of radioactive contaminates. But surely the repository's proximity to a series of working nuclear reactors is itself sufficient reason to give us pause. The ambition of eternal stasis that organizes the subterranean programme of geological repositories can do little to arrest the dynamism of life above the ground, as the comparatively short lifespan of the pyramids at Giza demonstrates. Moreover, if the accident is engineered *by design*, then the planning of ONKALO must assume that a nuclear mishap will occur at some point along the horizon-line of its future. There is, of course, no going back to a nuclear-free future in the sense that spent fuel and radioactive contaminates have been part of our collective planetary inheritance since the 1940s (natural sources of background radiation not withstanding). How to reinvent the nuclear accident is perhaps the greatest design challenge of all.

Planet

In the United States in the nineteenth century people believed if you farmed unsuitable land – dry land – rain would come as if by divine right: 'Rain follows the plow'. Throughout history we see similar folkloric ideas; water, it seems, can stem from the strangest things. Agriculture and supplications to a divinity might bring it, and notions linking rain to war are even older. They started with Plutarch, writing in ancient Greece that 'extraordinary rains generally follow great battles', and the ideas resurfaced in the US after the American Civil War (which was muddy indeed), where experiments abounded with guns and explosives, dynamite and balloons, all trying to produce precipitation.[1] The rains never came, although the US has been ever hopeful about manifesting them. After the Second World War, General Electric tested seeding clouds with silver iodide to make rain. Developed as a Cold War weapon, the storms caused by such cloud-seeding techniques reigned down (and rained down) during the Vietnam War, extending the monsoon season in the late 1960s. Now we get rain and war again; or, water and war and the hopes that the former might stop the latter. The process is subtler than sending up explosives and more peaceable. Now it is transpiration. Plants take up water through their roots and give it off almost like condensation, like breathing it into the air through their leaves. At least that is the idea: millions of trees will breathe water and return it to the ground, and the water will stop wars and the other disasters to come with climate change in the parts of Africa suffering from water scarcity. I hope the trees can do it.

The plan is for a belt of them across the continent. Called the Great Green Wall, it will spread nearly 7,775 km long and 15 km wide from Senegal to Djibouti and through 12 countries, including Mali, Niger, Nigeria and Somalia, some of Africa's most troubled places where millions are dependent on the land for their lives and livelihoods.

Thousands of miles away in London I think of this belt of trees on a cold rainy day in April. It is Earth Day 2016, and in Islington shoppers clutch paper bags sporting a green picture of the earth and announcing that H&M supports 'World Recycling Week'. The bags advertise recycling, while people buy new clothes. It is almost funny – the bags and recycling and fast fashion with its disposable ethos – but I am not sure whether to laugh or just be depressed. Around me people huddle and pull up the hoods on their parkas. It is the sort of weather a climate-change denier would take to heart, and a few days later it will even

Arguably part of the largest design project
on the planet, these tree saplings will be
planted as a buffer against desertification
in Africa.

Planet

Street Parade 1917

J. Sterling Morton, "Author of Arbor Day." Nebr City Nebr.

(Top)
An Arbor Day parade float, 1917,
including a bust of Arbor Day founder
J. Sterling Morton.

(Above)
J. Sterling Morton (1893–1897).

snow briefly. I shiver before H&M's windows emblazoned with three posters: one is the same image of the earth that is on the bags; the next exhorts me to 'Join In A Global Fashion Movement For The Planet'. The last promises: 'Recycling one single T-shirt can save 2,100 litres of water'. The repetition of 'one' and 'single' bothers me, as if the chain is struggling to drive its point home. The company's maths is also questionable. Cotton is among the most water-intensive of crops, and given that one new T-shirt requires 2,700 litres, maybe the disparity here is that recycling costs 600 litres? What if instead H&M just told people not to buy a T-shirt? Or, sent them to the Oxfam shop down the road?

A taxi behind me splashes a puddle onto the pavement, and here we are on Earth Day, 22 April, with a store creating its own 'global movement', and I am thinking of water and trees in Africa, while trees were the original point of the event. It began as Arbor Day in America in the nineteenth century, and it is also J. Sterling Morton's birthday.

Who? J. Sterling Morton? Largely obscure today, he had a walrus moustache and an upright bearing, and in 1872 founded Arbor Day. He was also a newspaperman in Nebraska like Charles Dana Wilber, who coined the 'rain follows the plow' dictum.[2] Both men practised a kind of Manifest Destiny via agriculture. Morton's idea was that the trees would bring rain and, more than that, civilization. In his thinking the head of man and the head of a tree were symbolically related (more on this below). The rain would come with the agriculture, and where pioneers would go, God would provide. It was a continent as a promised land.

Less than a decade after the Civil War had ended, the country was on the move and on the make. A transcontinental railway connected the coasts and the land in between needed settling. Morton believed trees were the way forward. He was a Democrat (which in those days meant a conservative). He had backed slavery in the Civil War and had tried his hand at politics. Frustrated in the pursuit, Morton built a farm and on it put a mansion, a replica of the White House, which might speak to his hubris and hopes.[3] Morton saw the Great Plains full of trees. His plan sounds like the Great Green Wall. He wanted them planted over this grassland, a prairie, which had been called the 'Great American Desert'. Now the aquifer below it is threatened and could turn the land into an actual desert.

The first Arbor Day had sponsored prizes: $25 in books for the individual who planted the most trees; $100 (around $2,200 in today's money) for the agricultural society with the highest numbers. More than a million trees were put in the ground that day. Farmer J.D. Smith singlehandedly tackled 35,000 of them. He took the $25 prize. I wonder what books he received? Did he think about the trees and pulp that had gone into them? By 1885 Arbor Day was a legal holiday in the state and commemorated across the country. Nebraska's governor, James W. Dawes, also moved the date to 22 April to mark Morton's birthday. That year his home town held a parade and a thousand students marched to the opera house. Two years later he gave a rousing speech at the state university linking trees to democracy and the Bible, as well as to the tree of knowledge and the tree of life.

'Ordinary holidays are retrospective', Morton said. 'Each of those reposes upon the past, while Arbor Day proposes for the future ... regarding it as an artist his canvas, and etches upon our prairies and plains gigantic groves and towering forests of waving trees, which shall for our posterity become consummate living pictures.'[4] He likened these to Peter Paul Rubens and talked about how Plato and Socrates taught under a grove in Athens. He went through the etymology of the word 'book', linking it to beech bark, and his linguistic analysis included how

Americans were descended from Druids. According to Morton, they planted the forests of England; 'Druid' apparently comes from the Greek *drus*, meaning oak.

For Morton the tree served as a moral lesson and a chance for some specious personification. 'Up in the clouds, gilded with sunshine, resplendent with coloring, nods the stately head; but down in the darkness and dirt are its supporters. ... Trees thus lead a dual life, an upper and a lower, so does man. The intellect, the reason, bathes in the light of knowledge ...' In the next paragraph, 'Man's intellectual life must dominate.' Oh, the will of man, just the repeated use of *man* and all that control and power Morton suggests ... pity the poor tree and the expectations put upon it. We still saddle trees with plenty of obligations beyond, say, transpiration or sending sap up from the roots in spring to the branches so they come into leaf, photosynthesize, produce nuts and fruit, sprout new trees and repeat the process.

Now in Africa the tree is salvation itself: peace, food and security. Here the 'Wall' stretches across the Sahel, which in Arabic means shore, as if this region were a sandy coastline. The coast, however, has no sea, just the Sahara at the Sahel's northern edge.

The idea for trees to keep the Sahara at bay has been around since the 1950s when a 'Green front' would stop the desert.[5] Plans reappeared in the 1980s, and in 2005 Nigeria's president, Olusegun Obasanjo, adopted the idea with hearty support from Senegal's president, Abdoulaye Wade, who reportedly coined the Great Green Wall name. The two presidents envisioned a literal line of millions of trees. Now the hope is that they will not just preserve water but also stop terrorism, and that just might be in the trees' power.

Kouloutan Coulibaly, Mali's national Director of Water and Forests, talked to the BBC about this in 2013. Trees could arrest the spiral of poverty that fuels extremism. They will return money to the economy by making land arable again. 'The green wall', he said, 'is an opportunity to provide jobs and combat poverty... it will develop these regions and I think it will be a solution [to terrorism].'[6] Another forester in the region said: 'The green wall is about giving people alternatives',[7] and Dennis Garrity, Drylands Ambassador for the UN Convention to Combat Desertification and Distinguished Senior Research Fellow at the World Agroforestry Centre, encourages western backing for these reasons. 'Pay a few cents now', he said to *The Nation* about the Green Wall, 'in order not to pay billions later. The Sahel region is approaching the verge of social explosion because of food insecurity and a lack of economic opportunity, especially for the young. There is recruitment for Al Qaeda going on right now. We need to provide a better alternative.'[8]

Who would not rather combat terrorism with a tree than a drone?

Aid is always in the eyes of the giver, framed in the issues of the day. The Belgian Congo, the harshest of colonial regimes, began in 1876 as Christian charity. Nearly a century later the US Agency for International Development (USAID) was going to stop communism. Started in the early 1960s by President John F. Kennedy, it provided foreign assistance, developing and supporting rural farming and electrical co-operatives because co-ops taught democratic principles of self-government. Now, trees and terrorism.

There is a contradiction, though, to trees in most of the Sahel. It could almost be a joke. Under the French colonial system that controlled most of the territory, if a tree was on your land, you did not own it. The state, the French, did. Trim it, you would be fined. Cut it down? Prison. You could not use the branches for firewood or anything else, so what could you do? Pull it up while it is still a seedling and pray nobody notices? These policies held in many places until just recently, and trees were relegated to tiny plantations that were ignored and failed. More wood was required to fence them in than they produced, meanwhile colonial powers had exported western farming practices to the region believing they would make the land more productive.[9] Monocultures, farmed in open fields that needed replanting each year, were the scientific way. Farmers sowed rows of crops that depleted the soil and ground water. With days of endless sunshine, the earth baked into clay or turned to dust and blew away. It is nearly impossible to picture windstorms that produce blizzards not of snow but soil as I stand on Upper Street in Islington and dream of summer.

One of the first people to notice these problems was a colonial forester, André Aubréville (1897–1982), who served in the First World War and afterwards applied to work in Africa because he wanted to do something worthy of war's sacrifices. (Aid is, indeed, in the eyes of the giver, and while it is hard to see the good in colonial administration, Aubréville believed in helping the world.) In pictures now he looks stiff. His hair is white, his glasses black and heavy. He wears the uniform of his job. In one image he is presented with a sword as he is inducted into the French Academy of Sciences, and he seems nervous, unsure how to pick up the weapon. It is inscribed with the forester's insignia of a horn and an African baobab flower. He has a military bearing, chest out, hair bristling in a brush cut that gives him the look of a hedgehog.[10]

He was the last gasp of colonialism and focused on tropical forests. He also coined the phrase 'desertification' that is used now to justify the Green Wall. The word described not just a desert but a process for becoming desert. Today 'desertification' has countless definitions, and it is easy to think it means places like the Sahara are spreading. They are not. The Sahara is a stable ecosystem. Instead, Aubréville was describing people's impact on the land and how we could bankrupt it, leaving it depleted, dry and un-farmable, as if the land were lost.

While Aubréville was in Africa, in Nebraska, home of Arbor Day, a desert grew. It was called the Dust Bowl. In the 1930s a swathe of the US from Canada to the Texas Panhandle experienced droughts. Industrialized farming had destroyed root systems of native grasses, and there were 'black blizzards' of dirt that blew all the way to New York. Nearly 800 million tonnes of soil were lost.[11] Farms, people, roads and homes were buried; the resulting mass migrations intensified the Great Depression. Desertification is not just something that happens over there on another continent far away. It happened in the US and helped create the American welfare state whose origins now often seem obscure.

In Africa Aubréville believed in close observation. His books were illustrated with hundreds of plates he drew himself. He brought botany's detailed studies to the forest, and did not see it as some Edenic place but as 'chaos and mystery'.[12] Forests were, he noted, never free from human influence, even going back hundreds of years.

His research focused as much on what was happening above the ground as below with root systems and the soil. In this I think also of J. Sterling Morton and his writing about the head of a tree, only Aubréville would hate that analogy and Morton too probably. The French forester called 'generalisations a danger to truth'.[13] He saw how fragile forests could be and the damage colonialism had wrought. It had caused over-planting, with burning back grasslands and brush, and the result: desertification. 'Who can say', he wrote in 1957, 'what might happen to the climate if the forests were to virtually disappear?'[14]

He also described how colonial administrators could, 'In just a few lines impose obligations and restrictions on an entire country, which the people concerned never even know about.' Aubréville did not necessarily envision a belt of trees as the solution but an understanding of ecology and human impact on it. 'The impression of desertification that one experiences before a savanna is that it has substituted a forest, which is not purely subjective and only somewhat sentimental.'[15] Those words – 'subjective' and 'somewhat sentimental' – stop me every time. We invest trees with so much beyond their power.

With Aubréville, 'desertification' enters the lexicon, and trees, which might seem the obvious solution, can make the condition worse. Their roots drink up precious resources from the water table. Morton had wanted exotic trees, ones that required copious water, planted on the plains.[16] I look at old pictures of Arbor Day. In 1887 12 women in elaborate dresses with trains and bustles gather before a handful of saplings on the prairie. Nebraska, 1901: a throng of school-children circle one tiny tree; a girl holds the shovel as if unsure what to do with the dirt. In a third photo, a brass band and people bearing a tree banner march down the street. Seeing the images I cannot help but think of what is to come. It is like seeing a ghost, seeing where the rain took the plough or vice versa because rains did not follow agriculture onto the prairies. In 1873, the year after the first Arbor Day, the US Congress instituted the Timber Culture Act. It guaranteed settlers in the Great Plains 160 acres (64.7 ha) free for the taking. All you had to do was plant 40 acres (16.2 ha) of trees. How many of them survived? How many of the million-plus planted on the first Arbor Day? Did J.D. Smith's? Morton's estate, Arbor Lodge, still boasts more than 200 tree species.[17]

It is easy to draw a line from Arbor Day to the Dust Bowl's desertification. Twentieth-century farming practices just made conditions worse. We had not come to a promised land. The Great Plains, the Great American Desert, often receives around 50 cm of rain a year with periodic droughts and high winds. During one wet cycle just before the Dust Bowl, farmers planted thirsty cotton with no cover crops on the fields in winter. Nothing held the soil. The solution to the black blizzards was something like the Great Green Wall. President Franklin D. Roosevelt deemed it a 'Shelterbelt'. Two hundred million trees were planted from the Dakotas through Nebraska, Kansas and Texas by the Works Progress Administration, part of the President's New Deal. The project was a success; only native species were planted, running along fences and boundary lines. China today has a similar project, the Three North Shelterbelt, which was begun in 1978 to repair over-farmed grasslands on the edge of the Gobi desert. The government has planted billions of trees but many wither and die. The favoured trees are species like poplar that require water but grow quickly to be harvested for pulp and paper.[18]

Along the Great Green Wall something else is happening, though, and the Wall might succeed because it is not planted with poplar but with native trees and brush. They are being allowed to regrow, and around them crops are planted. Called permaculture, the technique can look ancient when held up to western farming practices. An aid worker, Tony Rinaudo, accidentally reintroduced it. That story alone could make a fairy tale. This was in the drought of the 1980s that led to Live Aid, and he was so frustrated with seeding trees that blew away in dust storms that he tried something new, which happened to be very old. Rinaudo tied food assistance to farmers *not* cutting back brush, contrary to what western experts had advocated. He left the region in 1999 but a few farmers continued the practice, and it spread as those farmers' yields went up. In 2004 someone else found just how successful the idea had been. The US Geological Survey started mapping the changes and now 200 million trees flourish in Niger where Rinaudo had worked.[19] Hopes for the Wall look like Rinaudo's accidental

Schoolchildren in Omaha, Nebraska, plant
a tree for Arbor Day in 1901.

Planet

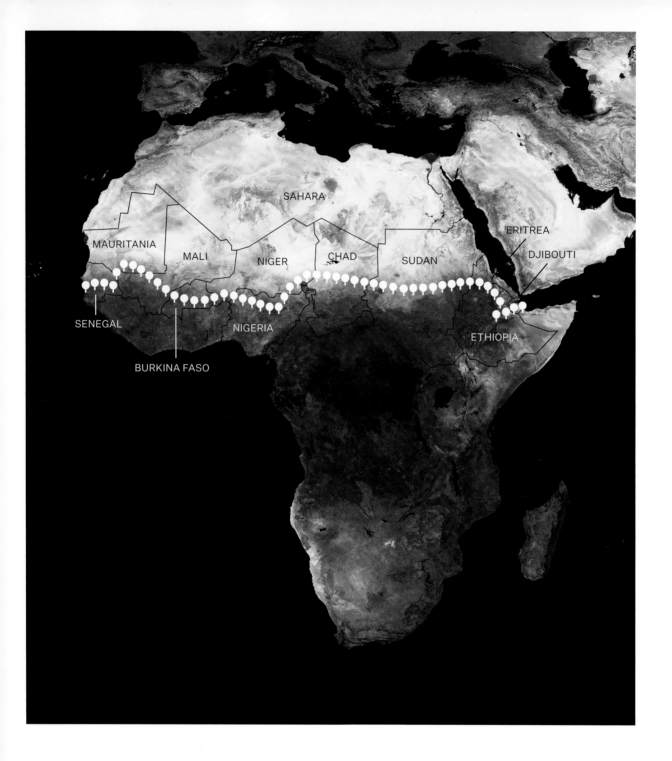

The 'wall of trees' to be planted across the
Sahel, marked by the white dotted line.

Planet

solution born of his frustrations. In the US the Shelterbelt is disappearing. You can see its remnants, a few stray trees, along highways. After the Dust Bowl a new kind of farming took over. In the 1950s intense irrigation gave farmers access to the aquifer beneath the Plains, and at current rates of use it will run dry sometime in this century. The aquifer will take 6,000 years to refill, while farmers cut old Shelterbelt trees seeing land that should be in production. Some of the techniques being taught to farmers to conserve water in the Plains look like those advocated by Rinaudo, and in China the World Bank has lent the government money to grow back the brush that had been considered a weed.[20] Instead of looking like the problem is over there, like Africa is distant and different, maybe it offers a solution.

When Morton first proposed Arbor Day, some on the State Board of Agriculture suggested calling it 'Sylvan Day'.[21] Based on the Latin for tree, 'sylvan' has other meanings – idyllic and peaceful – and I like that: a sylvan day. When the first Earth Day happened in 1970, its founders did not initially realize it was also Arbor Day. They were trying to think of a date that did not conflict with Easter and when students would not be on holiday. This Earth Day, while I stand in frigid Islington, at a meeting of the United Nations 175 countries sign the Paris Agreement on climate change. That too might offer hope.

Morton had declared trees were indiscriminate. 'There is no aristocracy in trees. They are not haughty.'[22] More than trees, it is poverty and drylands like the Sahel that do not discriminate. They cover more than 40 per cent of the globe and nearly half of all of Africa. There, 65 per cent of the population lives in places suffering from desertification and land degradation,[23] and people often get by on less than a dollar a day.[24] If the Great Green Wall can solve this, it does indeed seem sylvan. The expectations saddled upon the project, however, are vast. Just before Earth Day Nigeria's environment minister, Nkechi Isaac, said the Wall should be, 'Not just trees but economic trees'.[25] A couple of weeks later a conference dedicated to the Wall and desertification was held in Senegal. There, others talked about the trees stopping the refugee crisis, the West's latest worry. I want the Wall to. I want it all. If the land is arable, if people can live on it and support themselves, there is a hope for stable lives. Not long before I stood outside H&M the World Bank gave another $4 billion for the Wall.[26] The money is allocated for various projects, including folklore, hoping that oral histories and myths will reveal earlier farming techniques that have been ignored because so little is known about dryland farming and restoration. Meanwhile, ideas developed in the Sahel are spreading to the Caribbean, the US and beyond. At the conference Nkosazana Dlamini-Zuma, chair of the African Union Commission, said that if the Wall is a success, it will become one of the new wonders of the world – all that on the boughs of trees. The Great Green Wall is sylvan indeed.[27]

On 25 September 2014, NASA astronaut Barry E. 'Butch' Wilmore boarded a Russian Soyuz rocket. His destination was the International Space Station (ISS), his second time making the journey. Once there, Commander Wilmore would go on to perform four spacewalks, including one with fellow NASA astronaut, Reid Wiseman, to replace a failed voltage regulator. The trusty ISS toolbox they used, brought from Earth to aid such repairs, would have been filled with familiar items, such as various-sized pliers, cutters, screwdrivers and even a ratchet wrench.[1] And as so often happens on Earth, one of those tools sometimes goes missing. But what is there to be done in space, except sit tight (or float about) and wait for a replacement to make the long, expensive journey from Earth?

This was the very situation that Commander Wilmore found himself in when he lost his ratchet wrench on the ISS in late 2014. Fortunately, the astronaut had recently helped to install a zero-gravity 3D printer on board the space station. The first of its kind, the printer had been invented by the California-based company Made In Space, in partnership with the NASA Marshall Space Flight Center.[2] Hearing of Commander Wilmore's predicament, Made In Space swiftly put its newly installed technology – and its engineers – to work: 'We had overheard ISS Commander Barry Wilmore mention over the radio that he needed [a ratchet wrench], so we designed one in CAD and sent it up to him faster than a rocket ever could have. This is the first time we've ever "e-mailed" hardware to space.'[3] Just 34 years after the first object to be 3D printed (a cup about 5 cm tall) took months to produce on Earth, the entire process of creating

the replacement ratchet wrench – the on-Earth design, qualification and testing, and subsequent in-space 3D printing of the new tool – was completed by Made In Space in just four days. An incredible achievement, this innovation means that ISS astronauts no longer have to wait weeks, if not months, for spare parts from Earth. They now have the capability to make them instead.

I happen to have a collection of objects 3D printed by Made In Space on my desk. They are ground (or on-Earth) prints of pieces that were made on the ISS, including a version of Commander Wilmore's replacement ratchet wrench. At just under 12 cm in length, the wrench is easily the largest item of the set, the most recognizable in form and the most ergonomically sculpted. The remaining objects are much smaller in size. All, however, have been 3D printed – with their multiple layers easily seen – from a low-cost engineering plastic called ABS, a hard, tough and sturdy material with a long lifespan. But 3D printing with it can be difficult, even more so in space. 'There's no natural convection in a microgravity environment and heat is a huge thing for any 3D-printing', explains president of Made In Space, Andrew Rush. '3D-printing material is [also] often hydrocarbon-based and toxic in some ways when heated, so we had to ensure that we had systems in place to protect the astronauts.'[4] The ability for the zero-gravity 3D printer to filter toxic gases and nanoparticles, therefore, was one of the company's most important technical challenges.

More than just a convenience, the new technology has the potential to have an impact on the future of space travel more broadly, too. At present, for example, cargo and spare parts are launched

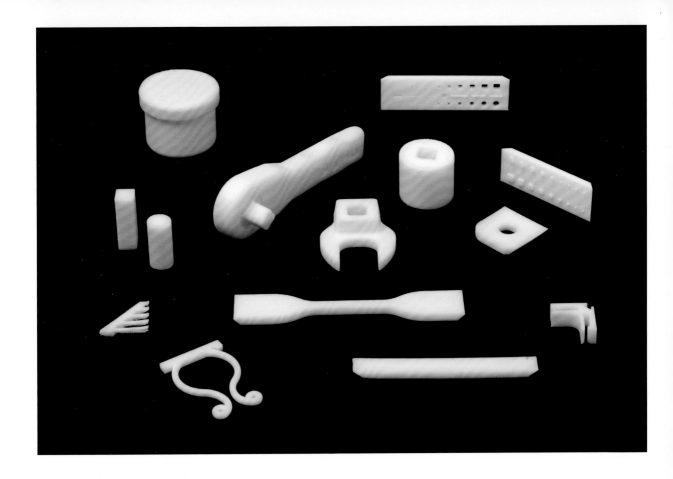

These objects were produced by a
zero-gravity 3D printer. What will it mean
to move our manufacturing to space?

Planet

via missions from Earth, which is not only an extremely costly process – NASA estimates that approximately $10,000 is required to launch just 0.45 kg (1 lb) of payload into orbit – but the objects launched must also be designed to fit inside or on a rocket, survive conditions during take-off (gravitational forces three to nine times those on Earth) and withstand extreme environments outside of our planet's atmosphere. In-space 3D printing eliminates many of these requirements, allowing instead for the creation of ultralight objects that use minimum material. Now more than ever, therefore, astronauts can take less with them, and when it comes to space travel, less is more: less cargo means cheaper and more efficient journeys into space.

Additionally, while Made In Space's printer operates with familiar materials like ABS plastic that would need to be replenished from Earth, future iterations of the technology have the potential to build from space-sourced alternatives. Washington-based asteroid-mining company Planetary Resources, together with its partner 3D Systems, for instance, revealed the very first object to be 3D printed with asteroid material at the Consumer Electronics Show in January 2016. A year earlier, architecture studio Foster + Partners released conceptual designs for a modular habitat for Mars created by 3D printing dwellings from regolith (the loose soil and rocks found on the planet's surface), and constructed by pre-programmed, semi-autonomous robots prior to the arrival of astronauts. Alongside Archinaut, Made In Space's latest venture, which entails upscaling their 3D-printing activities and establishing a factory in space, perhaps soon enough space travellers may not need to take very much, if anything, with them on their journeys from Earth.

These initiatives, which expand the possibilities of what we can achieve in space, are part of a much grander ambition: to journey to and colonize other planets, and Mars is our first stop. Identified by space researchers as potentially key to the future of the human race, Mars exploration has a great deal to offer, from new scientific knowledge, new resources and perhaps even a new (3D-printed) habitat. Theories around the possible presence of liquid water on Mars, furthermore, could lead to us establishing a more permanent home on the planet, an essential backup should we manage to destroy Earth. And under the current age of the Anthropocene, an epoch that recognizes the significance of human impact on the Earth's geology and ecosystems (including climate change), this

seems likely, if not inevitable: that is, if our scientific and technological progress has not already put humanity at risk of self-annihilation, as warned by world-renowned physicist Stephen Hawking.[5] The best we can do then is to prepare for the possible consequences, and inhabiting Mars is part of that preparation.

Getting there, however, is certain to be difficult, risky and hazardous. According to SpaceX founder Elon Musk, whose own company is determined to land the first human mission on Mars by 2025, the road will be 'dangerous and probably people will die – and they'll know that'.[6] Yet, the technology entrepreneur, who once famously declared that he himself wants to die on Mars, just not on impact, is confident that people will still sign up for the journey: 'They want to be the pioneers'.[7] It will be the bodies and brains of such people that will help us to reach the so-called Red Planet, individuals who are ambitious enough to sacrifice themselves for the bigger cause.

Meanwhile, for the rest of us who still need persuading to sign on to explore the dusty, desert-like planet, NASA has issued its own calls. In June 2016 the agency publicly released its 'Mars Explorers Wanted' poster series, originally a commission for the Kennedy Space Center from 2009. Each poster depicts astronauts performing a range of seemingly ordinary activities, from abseiling down cliffs to exploring canyons, accompanied by all-caps slogans such as 'MARS EXPLORERS WANTED' and 'TECHNICIANS WANTED'. The most propaganda-like variation is an appropriation of J.M. Flagg's Uncle Sam war poster of 1917. The NASA version portrays an orange Mars looming behind an astronaut pointing directly at the viewer; the words 'WE NEED YOU' appear just beneath. We need many things for our journey to Mars, this poster series seems to say, but most of all we need you.

The release of these images followed the public distribution of another NASA poster set called 'Visions of the Future'. Designed by the Jet Propulsion Laboratory (JPL), this series draws on the visual language employed by the Works Progress Administration (a US federal agency established in 1935) in its travel posters from the twentieth century, building on the approach to highlight the characteristics of a specific location in space: from Jupiter's auroras, reminiscent of Earth's Northern Lights, to the extrasolar planet Kepler-16b, with its striking double sunset. Adopting the concept of an 'exoplanet travel bureau', a somewhat tongue-in-cheek idea, to inspire wanderlust in

(Top)
Start-up Made In Space invented the first zero-g 3D Printer. In the near future, it wants to upscale its 3D-printing activities and establish a factory in space.

(Above)
Commander Wilmore shows off a ratchet wrench made with a 3D printer on the International Space Station.

Planet

viewers, the JPL team behind the poster series aimed to stimulate imagination: 'the genius of the idea [is that] these planets now are impossibly far away places', but to make it 'you need to dream'.[8]

Dreaming is certainly key. Originally intended for circulation within NASA only, these posters were created as a motivational tool for the scientists, engineers and other employees, reminding them of their long-term goal of reaching Mars and beyond. 'Someone needs to create the future, and you have to imagine that future if you want to create it', observe JPL Visual Strategists David Delgado and Dan Goods. 'Imagination is such a powerful vision of the future, way stronger than many other motivating factors', they add. 'If you have this beautiful vision of the future, people want to follow that and make it a reality'.[9] The posters in 'Visions of the Future', then, not only communicate the possibilities of space travel, but also remind us, quite literally, that the sky is (not) the limit.

Ultimately, these messages are devised to intrigue, excite and trigger imagination, and if we are to ever travel to and colonize other planets like Mars, this kind of enthusiasm, and the innovation it drives, is what we are going to need most. And although today it might take a leap to imagine how entire habitats for our life in space might be realized using our designed technologies, especially when it currently takes days to 3D print one ratchet wrench, we are only at the beginning of a much longer journey, one that many, including Hawking and Musk, believe to be the future of human civilization.

(Opposite)
'Imagination is our window into the future'.
A range of posters from NASA's Visions of the Future series.

(Above)
NASA's Mars Exploration Program seeks to understand whether Mars was, is, or can be a habitable world. This poster imagines a future day in Mars, and takes a nostalgic look back at the great imagined milestones of Mars exploration that will someday be celebrated as 'historic sites'.

Planet

AFTERLIFE

Current advancements in biotechnology and artificial intelligence have the potential to redefine our conceptions of what life is. Reawakening after death or uploading one's mind onto a computer are ideas that may sound like science fiction but are taken seriously by some futurists today. Alongside these efforts to preserve the self, other researchers are working to more generous aims, such as the preservation of genetic material, or even culture.

The Prospect of Immortality
Murray Ballard
2009

The Alcor Life Extension Foundation is an organization that freezes human bodies after death in the hope that technological advances may one day be able to reawaken them: a process better known as 'cryonics'. Here, a lorry delivers liquid nitrogen, the key ingredient for preserving bodies in pods located in the facility.

Afterlife

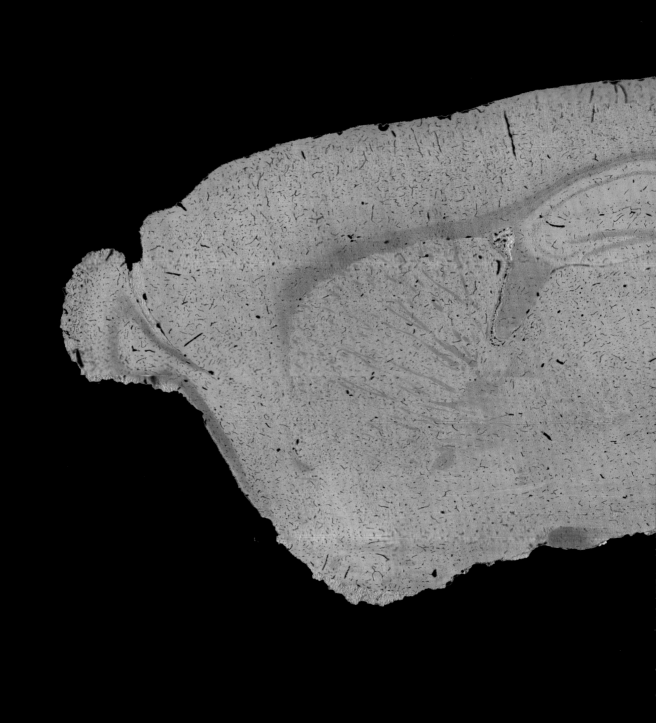

Whole Brain Emulation
3Scan
2018

This image captures an experimental process of brain scanning, used to create a digital model of a neural network. So far the technique has been trialled on small animals like mice (as shown here), whose brains contain fewer neurons than the human brain. 3Scan's larger aim is the transfer of information from a brain into a digital existence.

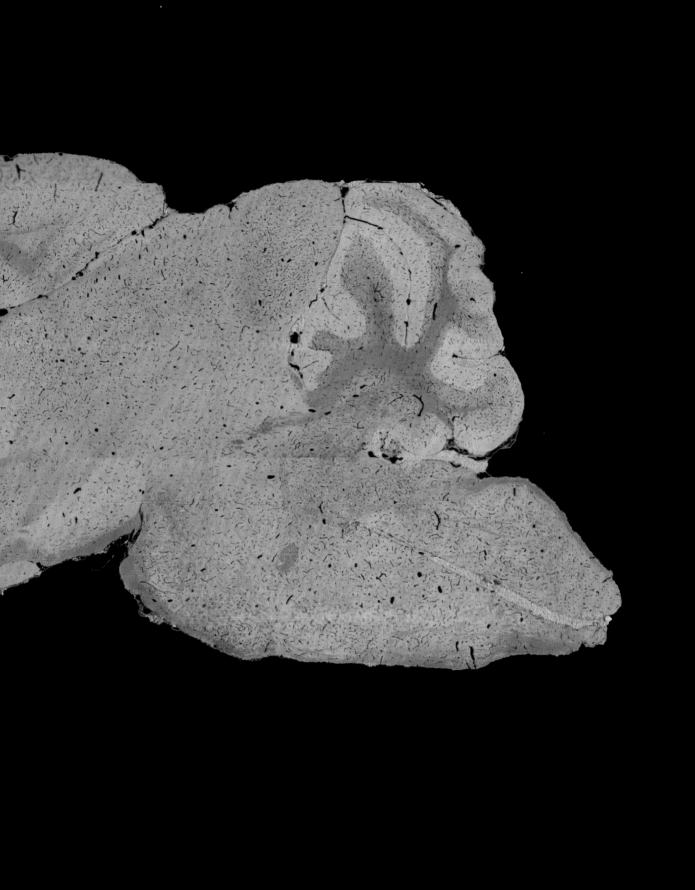

Afterlife

Why Will the Government Not Let
You Eat Superfish?
Alexi Hobbs
2014

Genetically modified salmon, like the one depicted, are seen by some as a solution
to decreasing stocks of wild fish, depleted by human demand. Although the prospect
of engineered fish entering our food supply is somewhat contentious, could the future
of our species rely on engineering the world around us?

Afterlife

Svalbard Global Seed Vault
Norwegian Ministry of Agriculture and
Food, NordGen and the Crop Trust
2008

Deep in an icy mountain in Svalbard, Norway, the Global Seed Vault is a large storage
facility built to preserve a variety of seeds from around the world. The initiative indicates
that perhaps the future lies not in protecting the human self, but in collective endeavours
like this, which involve conserving other species.

Afterlife

Imagine a language so small that you cannot see the order of its letters, made of a material that you cannot feel. It could tell you stories of the past and it promises whispers of futures. But to read this language – DNA, the code shared by all living things – requires lab protocols, computation and even courier shipments.

Our ability to read DNA has developed remarkably over the last 30 years. It took from 1990 until 2003 to sequence the three billion base pairs of the human genome, in a publicly funded effort costing $2.7 billion. By 2014, top-end refrigerator-sized sequencing machines could whizz through 16 genomes every three days, for around $1000 per genome.[1] As costs plummet and speeds increase, one outcome has been a 'democratization' of access to genetic data. But what does democratization mean here? Access to biological information is certainly not egalitarian. Since one sequencing machine from the market monopolist Illumina would cost $1 million (and they are sold in tens), most scientists still rely on centralized services to sequence DNA. They post samples off and wait a few weeks for the results to be returned.

A neat, grey box that fits in the palm of your hand, the MinION Mk1 is a pretty unremarkable-looking object. The most intriguing bit of its physical design is at the nanoscale, invisible to the naked eye. Five hundred and twelve pores made from hollow proteins, each just a couple of atoms wide, puncture a synthetic membrane with electrical current running across it. This tiny membrane is fixed inside a consumable cartridge – the flow cell – that clicks into the device. This little box is the world's first portable DNA sequencer with which

Oxford Nanopore, its maker, wants to democratize sequencing technology. If it manages to make DNA sequencing cheap and accessible, it would transform work in the biological sciences, but it could also impact our everyday lives, from healthcare to insurance to privacy.

To use the MinION, first you drop a prepared sample containing DNA molecules into the flow cell. DNA is like a string threaded with four kinds of bead, A, T, G or C: the four bases of DNA. Their precise sequence encodes genetic information. When the MinION is plugged into a computer, DNA is sucked through the pores. As the bases pass through a nanopore, each type disrupts the current across the pore differently, revealing the sequence (with the help of some computation). As code ticks out, it can be beamed into 'the cloud' to be matched against online DNA databases: you could identify a species from its sequences, or check bacteria for markers of antibiotic resistance. The device can run for minutes or days, until you have your answer.

Launched in 2015, the MinION's flow cells and software are regularly updated. In 2016, it had a higher error rate and was slower than the big sequencers, and so was better suited to smaller genomes like those of bacteria or viruses;[2] by 2017, the data rate increased tenfold and the MinION could sequence a much larger plant genome in a day in the field, and even read a human genome to more than 99.8% accuracy.[3] Meanwhile, the flowing ticker tape of information offers the tantalizing sensation of being able to 'see' the invisible molecules you are trying to read, rather than having to wait weeks to learn their identity. If the big sequencers

The MinION DNA sequencer, which fits in
the palm of your hand and costs about the
same as a smartphone.

Afterlife

are like mainframe computers, the MinION is more like the smartphone, the powerful, online computer in your pocket. Consumables aside, it already costs about the same as a smartphone.

It is this networked system that really begins to open up this little grey box's possible futures. The MinION has tested frog species in the Tanzanian jungle and Ebola victims in Guinea. There, the scientists tracked Ebola's spread by monitoring mutations in viral DNA sampled from infected patients.[4] Mapping an outbreak live means response times are reduced. The MinION is a node in a system of information and people and contexts. It operates in four dimensions: in space, over time. This portable, connected, low-cost sequencer could have far-reaching implications for how science is done, by whom and for whom.

Anyone can in theory buy a MinION for research use (except those in a few countries restricted by law).[5] But preparing samples continues to take time, money, equipment and scientific expertise. To address this issue, Oxford Nanopore is developing VolTRAX, a disposable device for processing samples; today, portable toolkits such as the lunch-box-sized, crowdfunded Bento Lab can already free scientists from the lab. However, you still need

Scientists in the Tanzanian jungle use the MinION to identify a new species of frog.

Afterlife

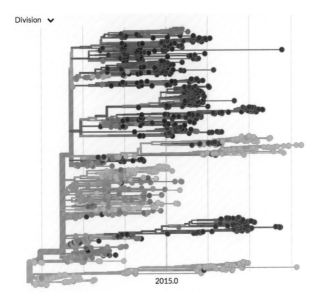

Division ⌄

2015.0

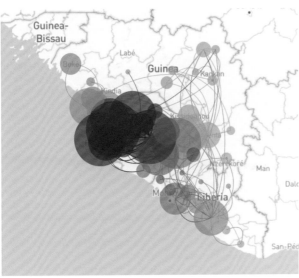

Genomic epidemiology of the 2013–16
West African Ebola Epidemic: phylogeny
(top) and transmissions.

chemical reagents and, for some applications, polymerase chain reaction (PCR) primers (sequences of DNA that are designed, and then printed or 'synthesized'). Both usually demand a lab address for delivery. Companies that synthesize DNA are heavily self-regulated for biosafety reasons, and screen every customer and sequence ordered.[6] The MinION may open up possibilities for science, but the technology is not yet accessible for all.

Oxford Nanopore says that it makes tools to measure data, not interpret it. Having data is not the same as knowing what it means, or what to do with it. As sequencing costs drop, consumer genomics companies have begun to promise democratization of access by sequencing your data for you, and interpreting it too. The privately held biotechnology company 23andMe has offered a personal genome service since 2007. Send off a vial of spit and 23andMe would sequence a fraction of your genome to reveal details such as ancestry and earwax type, and report on serious genetic risk factors including for Alzheimer's disease and hereditary breast cancer.[7] In 2013, however, the US Food and Drug Administration (FDA) banned 23andMe's $99 test on the grounds that anything marketed with the intention of preventing or diagnosing a disease must be regulated as a 'medical device'.[8] 23andMe launched a pared-down product and pitch two years later; in 2017, the FDA permitted risk analysis for 10 diseases.[9] In the United Kingdom, customers have been receiving this sensitive information, without genetic counselling, all along.

Despite such regulatory bumps, the commercial push to sequence the public is growing. In 2015, the influential biotechnologist J. Craig Venter launched Human Longevity, Inc., and partnered with a South African insurer to subsidize the sequencing of a million insurance customers' exomes (a useful two per cent of the genome). The world's largest archive of human medical and genetic information – plus insurance information – will soon be in private hands.[10] Whole genome sequencing for consumers is becoming viable as well. Veritas Genetics, a spin-off from Harvard University's Personal Genome Project, started to offer the service in 2016 for $999, including interpretation and genetic counselling.[11] It markets the product to consumers, but avoids regulatory issues by insisting your doctor orders the test.

Public sequencing projects are being dwarfed by companies selling subsidized tests to facilitate the mass private accumulation of valuable genetic information, mined for research and sold

Afterlife

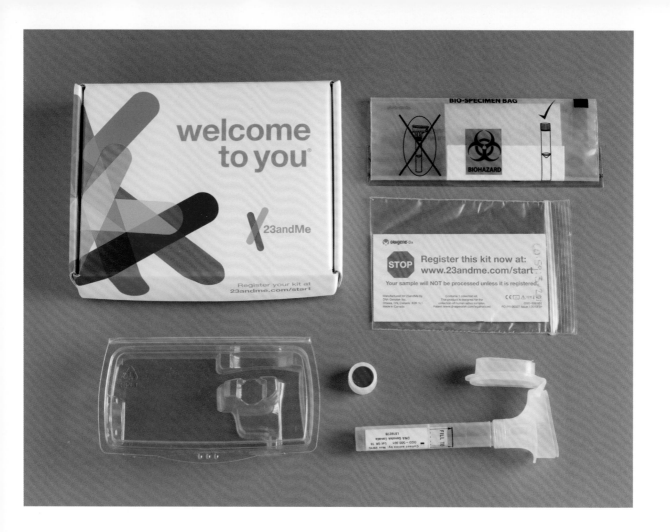

This 23andMe saliva collection kit enables users to access their genetic information: order online, spit, send and discover – but perhaps at a cost of privacy.

Afterlife

('de-identified') to third parties.[12] 23andMe has disclosed that it has been approached by law-enforcement agencies for customer DNA, requests that the company claims to have denied.[13] Crucially, if you have not volunteered your DNA and your relatives have, these companies still know plenty about you. To keep their data secure, the consumer has to trust that companies will remain ethical, and in business, in perpetuity.

Veritas's slogan is 'Live in the know™'. But more testing does not necessarily mean better health, and genes are just one, complex part of your biological story. Simply weighing yourself is a better indicator of your future health.[14] Yet Veritas's website says that 'someday soon, we predict we'll all be carrying full genome cards and using them to select our food, personal care products, and fitness routines'.[15] Consumers are throwing their genomes at the architects of the consumer genetics industry, without sufficient genetic or legal literacy. The market is shaping the technology: is this what we want from democratization?

As the story of 23andMe shows, regulation is lagging behind technological advances. In 2008, the USA introduced the Genetic Information Nondiscrimination Act (GINA) to protect access to health insurance and employment, but the UK has no such law.[16] Today, you could publish your genome online without repercussion. But as the industry grows, will insurance companies and personal genomics firms lobby for regulation to match their commercial interests?

Oxford Nanopore may just make tools, but people with a variety of agendas will use them. After decades of public investment, the market picked up on personal genomics. Now, the mobile sequencer is poised to seed new industries: Oxford Nanopore sees not just democratization, but also ubiquity as the goal. It wants 'to enable the analysis of any living thing, by any user, in any environment'.[17] Networks of nanopore sensors could sequence living systems in real time, beaming data to Oxford Nanopore's offshoot, Metrichor, for cloud-based analysis. Our privacy, our behaviours and even our identity could be challenged by future iterations of this small grey box. Welcome to what Oxford Nanopore calls the 'internet of living things'.[18]

Food producers could verify provenance (is your burger really beef?), farmers could survey crop health and pathogens could be tracked on the Tube. You could even monitor yourself. Your blood is flowing with information, as dead cells shed DNA into the bloodstream. Prenatal screening already uses foetal DNA sampled from the mother's blood; since cancerous cells grow and die fast, Illumina is now developing liquid biopsies that could catch signs of cancer earlier.[19]

But 'self-quantification' is Oxford Nanopore's target, rather than testing services. It imagines consumer devices that monitor markers in the blood every day, tracking changes and spotting trends. If these devices do not give diagnostic advice, they would not need medical approval, opening new markets before regulations catch up.[20] Streaming live data from millions of us, and from our environments, could be collated, mined like the genomics companies' databases, to learn more about our health and our habits. Who would own this data and what would it be used for apart from public health?

Predicting futures from an existing object is difficult, but an object can make certain futures more likely, as those futures are normalized in our imaginations. The MinION exists, but the internet of living things does not. Do we want it? No future is inevitable, even if we are told that it is. As with any technology, the MinION is shaped by existing values and it is our values that shape the future, not technology itself. Democratizing sequencing seems to be a good value. But we should ask: democratization to whose ends? This little grey box should certainly persuade more of us to read the small print.

Afterlife

In April 1773 Benjamin Franklin wrote a letter to French physician and botanist, Jacques Barbeu-Dubourg, in which he discussed methods of bringing back to life people apparently killed by lightning. Noticing that some animals and plants undergo suspended animation, and (probably erroneously) recounting how flies apparently drowned in Madeira wine before being shipped across the Atlantic could sometimes be revived by the rays of the sun, he suggested:

> I wish it were possible, from this instance, to invent a method
> of embalming drowned persons in such a manner that they may be
> recalled to life at any period, however distant; for having a very
> ardent desire to see and observe the state of America a hundred years
> hence, I should prefer to any ordinary death the being immersed
> in a cask of Madeira wine with a few friends till that time, to be then
> recalled to life by the solar warmth of my dear country![1]

But he concluded that in all probability he lived 'in an age too early and too near the infancy of science' to see this achieved.

Once upon a time people who did not breathe were regarded as dead. By the 1700s, however, physicians discovered methods of first aid that brought apparently drowned people back to life. The absence of a pulse was still a reliable sign of death, until chest compression and defibrillation began to be practised in the twentieth century. Now death became a matter of irreversible cessation of brain function, even if the rest of the body was functioning well.

As medicine advances people who were previously considered as beyond saving or dead have become merely critically ill. This means that many people who today are regarded as dying may in the future be in a salvageable state. If only they could somehow end up in that future they would have a chance.

CRYONICS

Cryonics is based on this idea. What if a terminally ill patient could be placed in a stable state so that their condition at least did not get worse, awaiting the time when technology is good enough to save them?

In the case of cryonics the approach is to cool down the body as soon as possible after legal death to extremely low temperatures – so low that no

The author's 'Cryonics Alert' bracelet, which gives instructions to medical teams in case of death, such as 'Do CPR while cooling with ice'.

Afterlife

'Cryonics Alert' bracelet, front.

Afterlife

metabolic or chemical change will occur. By infusing with protective chemicals, damaging ice crystal formation can be prevented, turning the water into a glass. Eventually the body (or just the head in the case of us cheapskates who seek only to preserve our brains) will be stored in a tank filled with liquid nitrogen at -196°C.

The concept hinges on two key assumptions. Firstly, that the damage done by the preservation process and clinical death that we cannot currently reverse will become reversible sometime in the future. Secondly, that being legally dead does not imply irreversibly dead: tissues become non-viable on a timescale of hours, so a quick intervention may prevent further deterioration.

Hence the need for medical alert bracelets to make the emergency room doctors aware that this patient needs to be treated somewhat differently. Normally, legal death is when medicine slows down; here, legal death is the signal for a frenzy of treatment aimed at hopefully stopping real death. Once the suspension is well underway things slow down again, eventually approaching the perfect stillness of storage in liquid nitrogen at a temperature where no chemical reactions or decay happens on timescales of centuries.

RATIONALITY

Cryonics is interesting as a subject because it forces us to think about our expectations of the future and future technology to decide whether or not it is worthwhile.

Someone who expects future technology to be like the present – perhaps we have already reached the ultimate limits of technology? – would not be interested. A pessimist who anticipates that the future will be dark has no reason to sign up. Nor does someone who foresees disaster, whether an end of the world scenario or merely commonplace disruption during their suspension.

Conversely, if one has a more positive outlook on technology and humanity's overall chances, cryonics can make sense. There are still issues of personal identity and assumptions about what technologies will be available. Will future medicine work by repairing the body cell by cell, so that the rational choice is just to ensure that the suspension does as little biological damage as possible? Or would it be a matter of analysing the brain digitally and reconstructing it more or less from scratch, either biologically or as software in a computer – in which case it might be more important to preserve information rather than biological viability?

It is all a gamble, but given that a dying patient has little to lose, it is a rational gamble for many.

CONFRONTING MORTALITY

As a way of escaping death, cryonics is not very reassuring. It forces you to confront your biological mortality head-on: signing over your corpse as a medical donation, filling in forms about what to do if preservation conditions are less than perfect (what organs should they save?), and accepting that this is an extremely experimental medical procedure that might be performed under less-than-ideal conditions.

This might explain why so few people are signed up (about 2,000 worldwide) despite the promise of vastly extended life. Numerous people cling to notions that reassure them about their afterlives, be they healthcare reforms, religious salvation or the solidity of a British countryside cemetery. Cryonics' acknowledgment of its uncertainty and fallibility is honest but not very appealing.

An American Index of the Hidden and Unfamiliar, 2007
Taryn Simon

Framed archival inkjet print and Letraset on wall
94.6 x 113.7 cm (37 ¼ x 44 ¾ inches)

This cryopreservation unit, at the Cryonics Institute, Clinton Township, Michigan, holds the bodies of Rhea and Elaine Ettinger, the mother and first wife of cryonics pioneer, Robert Ettinger. Robert, author of *The Prospect of Immortality* and *Man into Superman* is still alive.

The Cryonics Institute offers cryostasis (freezing) services for individuals and pets upon death. Cryostasis is practised with the hope that lives will ultimately be extended through future developments in science, technology and medicine. When, and if, these developments occur, Institute members hope to awake to an extended life in good health, free from disease or the ageing process. Cryostasis must begin immediately upon legal death. A person or pet is infused with ice-preventive substances and quickly cooled to a temperature where physical decay virtually stops. The Cryonics Institute charges $28,000 for cryostasis if it is planned well in advance of legal death and $35,000 on shorter notice.

Afterlife

The concept of cryonics is fairly well known, but most people have never met anybody who is signed up for it and often assume it is just science fiction. I wear my cryonics alert tag openly, which leads to interesting conversations. Often the questions turn to how I might find myself in a dystopian future: there seems to be a psychological need among many interlocutors to find arguments why cryonics is not rational, a need we less often see about views on religion or health (how many believers are asked what they would do if heaven turns out to be dystopian?). By framing itself as an attempt at rationality cryonics opens itself up for criticism that is rarely applied to belief.

WHEN THE FUTURE INTRUDES INTO THE PRESENT

Cryonics also allows the future to intrude into the present in a slightly unusual way. We often think about the world in a 'near mode' of things that affect us – our families, mortgages, our actual environment – and a 'far mode' of things beyond – technology, what happens in the media, the Future. These are frequently inconsistent: many surveys find that people expect a fairly dark future in the large – climate change, inequality, alienation – but have very optimistic views about their own future with friends, family and prosperity.

Cryonics means bringing the far into the near. My alert tag indicates that I can be thrust into the future at any point. Paying the life insurance that will fund my possible future suspension means I am buying a ticket to the future. Many of the conversations I have about the possibility dwell on how my social bonds will be affected by a cryosuspension. When I married, the priest saw my cryonic tag and my vow hence ended '... till death do us part (temporarily)'.

Another effect of cryonics is that distant risk becomes near risk. To a suspended person risks from corporate bankruptcy, societal changes and existential risk to humanity become compressed into what is essentially a medical risk. To reduce this, it may be wise for us to make the world a safer and more stable place. Working towards that goal may also be a good way of ensuring the future will want to revive us.

To our senses Benjamin Franklin's sun-warmed Madeira wine is far from the icy liquid nitrogen of cryonics. But conceptually they are the same: a possible way of escaping certain death in the present, for an uncertain future at the limits of our understanding, but one filled with optimism. Maybe we are living in the infancy of science, but not too early.

In 1950 the British computer scientist and cryptographer Alan Turing posed a simple yet provocative question: 'Can machines think?' Written at a time when computers were clunky, single-purpose and took up entire rooms, Turing anticipated – and set in train – the core themes of artificial intelligence (AI) research to this day. Leaping effortlessly across mathematics, logic, philosophy and religion, he pointed towards a future where a machine would be able to imitate a human, to pass what would become known as the 'Turing Test'. He foresaw a universal computer, subtle, sophisticated, intelligent, knowing when to follow the rules and when to break them. The way to create such a machine, Turing proposed, was not by 'trying to produce a programme to simulate the adult mind', but 'rather to try to produce one that simulates the child's'. If this child mind were subjected to an appropriate course of education, Turing continued, 'one would obtain the adult brain'.[1]

In the years since Turing's proposition, our child machines have been growing up. They have learned to speak and listen; to see and recognize faces; and to walk, run and help around the house with simple tasks. We have put them to work in factories on the production line; in supermarkets as digital checkout staff; in call centres fielding customer service queries; and in journalism, writing basic news stories. Having clocked millions of kilometres behind the wheel, machines are now even learning to drive cars. I like to think we are currently living in the 'teenage years' of AI: our machines are learning to function independently in the world, but they are still living at home under the protective wing of their human creators. Now, like any other teenager, our machines are learning to play video games.

In February 2015, DeepMind, a subsidiary of Google, published research on how it had trained a computer 'agent' to play various games on the Atari 2600 console. Using only the pixels and game score as inputs, DeepMind's agent was able to learn to play early 1980s classics such as Space Invaders, Breakout, Robotank and Pong after only a few hours of practice. In explaining its achievement in the scientific journal *Nature*, DeepMind writes: 'we developed a novel agent, a deep Q-network (DQN), which is able to combine reinforcement learning with a class of artificial neural network known as deep neural networks.'[2] I am not going to pretend to know what all this means – the paper is extremely technical, intended for expert computer science peers, not mere curators – but essentially it has created an algorithm that is able to improve at a task by rewriting itself, much like the human brain.

DeepMind's agent begins by randomly moving and hitting the fire button in an attempt to score some points. When it does, it remembers what it did, in order to replicate this action next time. When playing Breakout, for instance, the machine is pretty rubbish for the first hour or so, but steadily improves as it racks up experience. After four hours of practice, the machine consistently scores better than a professional human player.

What is the point of a computer that can play video games? DeepMind is not motivated by claiming the highest scores in the pinball parlour, but by creating tools that can think and learn like humans. AI is held up to be the answer to the world's most

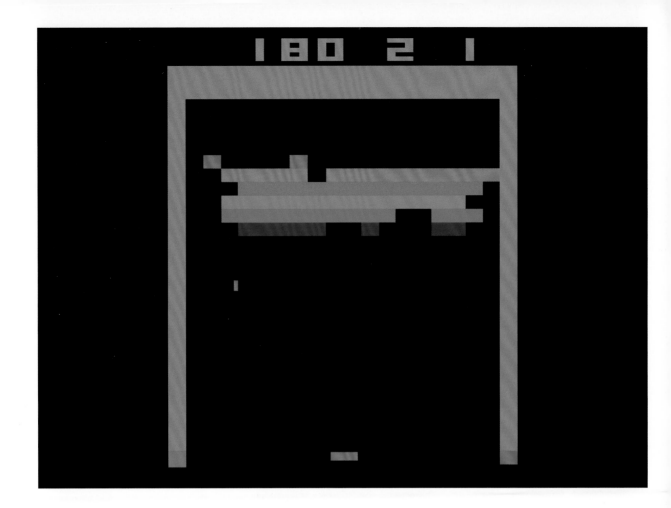

After only a few hours' practice,
DeepMind's artificial intelligence could
play the 1980s Atari game Breakout better
than a professional human player.

Afterlife

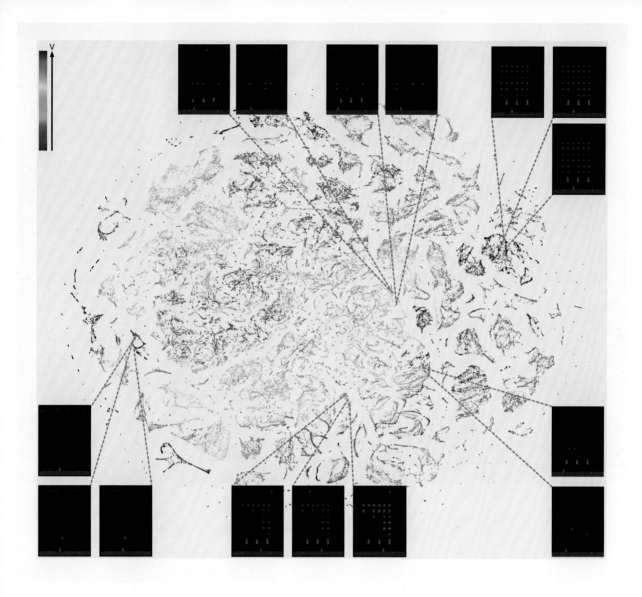

A diagram of a meta-analysis of the
learning system, showing anticipated
scores in Space Invaders.

intractable problems, from filtering spam to solving climate change. As DeepMind founder Demis Hassabis explains, its DQN 'is really just the first baby step toward this grander goal, it's the first example of a full system that can actually learn to master a wide range of diverse tasks.'[3] It is this ability to play *different* games that is the real breakthrough here. Earlier game-playing robots, such as IBM's Deep Blue that famously defeated chess grandmaster Gary Kasparov in 1996, were intentionally designed to play one game and one game only. DeepMind's DQN agent can master a variety of games straight out of the box, without any prior knowledge. As AI pioneer Professor Stuart Russell explains: 'If your newborn baby did that you would think it was possessed.'[4]

As it happens, we have got a baby at home. While she is no good at playing video games, it is fascinating to watch her learn. Babies are driven by a relentless curiosity for making sense of the world. They test it, push it, tip it, lick it, drink it. With every shake of a maraca or taste of yoghurt, you can practically see the neural pathways being laid down in front of your eyes. Watching videos of the DeepMind agent learning, I recognize a glimmer of this same curiosity and experimentation. Of course, it is very human of me to anthropomorphize the machine; it is not 'curious', but merely programmed to achieve a high score through trial and error. Yet seeing something improve through the accumulation of experience is a trait I am only familiar seeing in living things, like children and pets, not machines. Watching DeepMind in action, the answer to Turing's question, 'can machines think?', is surely 'yes'.

The machines are not very clever yet ('Siri, call mum', 'Playing U2'). We can still laugh at them and maintain our smug superiority. Although it may have taken thousands of researchers many decades to nurture Turing's child robot into a teenager, algorithms that can undertake sustained self-improvement could presumably grow up very fast, bootstrapping themselves to become far more intelligent than even the brightest of human minds. While the capacity of our brains is limited by the size of our skulls, there are few practical limits on the capacity of a disembodied digital brain. Connected to all the information on the internet, it could become very smart very quickly. Nick Bostrom, of Oxford University's Future of Humanity Institute, describes this as an 'intelligence explosion', where an AI could advance so rapidly (and seize

DeepMind's artificial intelligence defeated world champion Go player Lee Se-dol in March 2016, a game long thought to be impossible for computers to play at the highest level.

Afterlife

control of our systems) before its human designers even realize.[5]

As the creators of these machines, it is naturally presumed that we are the parents, not the children. But what if it is the other way around? 'Before the prospect of an intelligence explosion', writes Bostrom, 'we humans are like small children playing with a bomb. Such is the mismatch between the power of our plaything and the immaturity of our conduct.'[6] By recasting us humans as the naive ones, rather than those with all the answers, Bostrom calls for humility and caution in our dealings with AI. With so much to be gained from having such an artificial brain under one's control, there is plenty of effort being put into creating it, but little exploration of the possible negative implications. We can hope that Google and DeepMind's intentions are pure, and that they are taking the necessary precautions, but is that enough? We do not yet know what shape a superintelligent AI will take, let alone how to design the box to contain it.

On a recent trip to San Francisco I had lunch with Stuart Russell, author of the definitive textbook on AI, founder of the Centre for Intelligent Systems at UC Berkeley and self-appointed social conscience of AI research. We spoke primarily about the 'laundry robot' that he and his team are developing.[7] While laundry might be something we do mostly without thinking, it is a challenge for machines, comprising a number of sequential steps in a messy real-world context. It is, therefore, the perfect training ground for the next generation of AI. Despite the unthreatening nature of the laundry robot – it takes 15 minutes to fold a towel – Russell is equally concerned with what happens when these machines do become properly smart, asking the simple yet provocative question: 'What if we succeed?' In an open letter to the AI research community, Russell states that 'we need to put a lot more thought than we are doing into what the precise shape of this [intelligence explosion] event might be.'[8] Backed by more than 8,600 signatories, including the founders of DeepMind, the letter clearly touched a nerve. When Steven Hawking or Elon Musk say that AI could spell the end of the human race, they are channelling Russell.

Despite these warnings, Russell is ultimately optimistic, advocating for the 'societal benefit of AI'.[9] A superbrain applied to the public good could eradicate disease, poverty, global conflict and climate change. Who would say no to that? For this bright future to come to pass, Russell

argues, it will require AI to be 'provably aligned with human values'.[10] Our perception and understanding of the world has evolved over millions of years. Can we expect a robot to share these values straight out of the box? As Russell writes:

> If you want to have a domestic robot in your house, it has to share a pretty good cross-section of human values; otherwise it's going to do pretty stupid things, like put the cat in the oven for dinner because there's no food in the fridge and the kids are hungry.[11]

This new oracle, a parental figure with all the right answers, has the potential to save the world, but in order to do so we need to learn to trust it.

Rather than being a threat to humanity, the advent of superintelligence presents a chance to remake society: a world without work, disease, famine or war, where all our needs and desires are taken care of. Instead of forced retirement, it will be a return to childhood: all watched over by machines of loving grace, subsumed in the warm bosom of our computers' infinite wisdom and love. Whether AI ushers in this public-spirited utopia, or spells the end of humanity, is ultimately up to us.

Afterlife

FUTURES IN THE MAKING

Arjun Appadurai

This volume and the exhibition to which it is connected are dedicated to identifying the seeds of the future in the design practices and innovations of the present. They call our attention to objects and forms that seem to mark our times and to contain the outlines of our material futures. This brief chapter seeks to step back from the rich examples that are curated and presented in this exhibition and ask what they might demand from us as users, viewers, citizens and ordinary human beings.

NEW TOOLS, NEW WORDS

One remarkable feature of this new material world is that it is accompanied by a change in our linguistic world. Many words from our past have been repurposed to signal different meanings and these linguistic adjustments themselves constitute a map of our changing times. Examples include ordinary words such as 'mining', 'code', 'user', 'printing', 'searching' and 'liking'. Mining is a key word in the world of big data, and refers to the technologies and protocols utilized to scan huge data sets to discern patterns of behaviour, preference and interest among large populations. That one of the words to describe this method of pattern recognition is the one that humans have long employed for the extraction of precious metals from the earth's depths, points us to the fact that 'mining' still has to do with the unearthing of value by extracting, refining and sifting through raw materials, and continues to involve putting human misery at the service of profit. 'Data mining', the form of mining that is most vital to our emerging technologies, is indeed both exploitative and profit-oriented but the technologies through which it works appear often to be clean, transparent and democratic. Yet the persistent use of the word 'mining' should keep us alert to its older meaning. 'Code' is another such word. Once used for any linguistic form that underlies another one, it gradually began to be associated with secrecy, encryption, warfare and espionage. Today, across subjects such as computer science, genetics, cryptography and programming in general, it has acquired a neutral, professional and benign quality, but the truth is that all 'coding' involves specialized knowledge (of languages, algorithms and protocols) that is intended to be invisible to the average user, unavailable to competitors and resistant to 'cracking' or 'hacking'. So codes and coding are also becoming a part of our linguistic environment but their new meanings may hide older problems and risks. 'User' is even more commonplace in our emerging linguistic worlds and it is another word that had simpler meanings in the past, and referred to certain forms of value ('use value') and to the basic role of utility in human life. But today we are all 'users' of tools, apps and devices that define much of our humanity and sociality. Here again, what this linguistic change indicates is the colonization of a significant amount of our everyday existence by rules of function, utility and convenience that have gradually begun to define our views of our lives in greater ways. Our cities, our institutions, our politics, our policies and our collective values are increasingly defined by such qualities as smartness, speed and connectivity rather than by values such as inclusiveness, equity or justice. The everyday habit of regarding ourselves as users is the linguistic edge of this process.

It is easy to see how the technologically induced transformation of other words, such as 'printing' (to describe 3D reproduction), 'searching' (as defined by Google) and 'liking' (as normalized by Facebook), follows a similar logic in which old meanings are both invoked and subverted by new devices and protocols. This logic can also be extended to such basic terms as 'identity', now often considered as a vulnerable form of intellectual property that can be stolen, and 'privacy', now mainly viewed as a target of machine surveillance.

These linguistic examples can be multiplied and they are in some ways a natural part of all periods of rapid technological change. But in our times, new meanings are so quickly circulated and naturalized as to make them seem minor, even though they have fundamental implications for our humanity, our sociality and our solidarity. Such linguistic adaptations are part and parcel of subtle cultural shifts with regard to scale, visibility and individuality.

THE TRANSFORMATION OF SCALE

The future emerging through design demands a new way of thinking about scale in our lives as human beings. Scale has always been an important dimension of human life, and in earlier eras smallness and bigness had predictable relationships with regard to increasing size, distance, connectivity and materiality. Above all, the scales of time and space had a relatively hierarchical connection in which nearness in space and time meant greater access and familiarity and remoteness meant less knowledge and control. That hierarchy is now no longer something we can take for granted. Scanning, sensing, screening and reading have made the scales irregular, and smallness and nearness have collapsed into more obscure qualities due to the ubiquity of mediation and the saturation of our intimate spaces by big data, electronic connectivity and mobile applications. Telephony, mobility and multi-use personal devices have diminished the boundaries between our bodies and our applications, and devices for self-surveillance (of our location, our health, our fitness and our finances) have turned our bodies increasingly into nodes in a complex, large-scale communication and data-gathering universe. At the same time, developments in robotics and biomedicine, along with various social applications, have extended our bodies through prostheses, proxies and electronic personae so as to allow them to extend themselves remotely. These developments have upset the traditional relationship between small and big, near and far, here and there, so that scales are now cross-linked outside the architecture of hierarchical orders.

VISIBILITY AND INVISIBILITY

Although the professional and ideological worlds of contemporary design are aggressively secular, they are marked by utopian urges and thus share something with the ethos of religion. Among other things, religion has always been about how to understand and manage the relationship between the visible and the invisible. Human beings are aware that their lives are deeply affected by invisible forces. Cosmologies and theologies seek to account for the ways that these forces shape reality. Magic and ritual help human beings to negotiate this boundary.

The future that is emerging through our new designs changes the character of the negotiation between the invisible and visible orders. One site of this negotiation is the screen: on our laptops, our Kindles, our iPhones and iPads. These screens now set the conditions in which we shape our lives as users, as citizens, as consumers and as friends. We rely on our screens to keep us in social contact, to learn about the news, the weather, the stock market, and to play games, listen to music and order taxis. The screen as an interface defines the site where the invisible becomes visible. As we develop innovative platforms, servers, tools and apps, the interface is where the world is made visible, manageable and available to us. And as this interface grows more complex and ubiquitous, we adapt to the landscape of 'clouds', data 'mining', information 'farms' and the like, and start to regard them as new forms of nature, which are also becoming second nature to us. As we play, socialize and dream through the interface, novel ideas of wellbeing, uncertainty, community and humanity are gradually beginning to be part of our common sense. From this perspective, the future emerging through design is a sort of religious future, although it appears in the guise of technology.

INDIVIDUALITY

Most striking about the new materialities that shape the design of our emerging futures is that they may be changing the very bedrock of western modernity – the category of the 'individual', that package of agency, value, motivation, personality and choice that underlies our most common and taken for granted ideas about knowledge, justice, freedom and sociality. Other societies in human history have had different ways to see the distribution of agency and personhood in the cosmos and have regarded human beings as only one crystallization, among many, that define agentive diversity. Often, human societies have viewed themselves as 'dividuals', that is, temporary assemblages of matter and meaning, quick to change in the process of interacting with other 'dividuals' (human or not) in an unstable universe of flows and exchanges. This sort of 'dividual' may be emerging again in our times, with machine protocols that divide us into attributes of wealth, race, location, taste or class so as to better determine our suitability for credit, insurance, educational assistance, welfare or refugee status. Contemporary design also does not rely much on the classical individual, seen as a permanent repository of rights, capacities and obligations, but is mostly oriented to dividuals, clusters of tastes, locations and biographies that can be combined and recombined to create greater aggregations or patterns for the purposes of markets, states or other large agencies. Thus, inevitably, the term 'social' is also changing its meaning, and no longer indicates contractually organized, permanent and well-defined groupings of individuals, but is now a constantly shifting array of dividuals, who can briefly connect through technological devices, but who are no longer easy to unite in terms of such older categories as class, community, locality or ethnicity.

I began by indicating that words are acquiring new meanings in our technology-driven and designed futures. I also suggested that this linguistic change is the symptom of a deeper set of cultural changes in such areas as scale, visibility and individuality. I do not mean to be alarmist or to indicate that we are entering some sort of design dystopia. Rather, I suggest alertness to the tectonic shifts beneath our modern conveniences, affordances and utilities. Without such alertness we will become victims of the volatility of the world of designed futures rather than its masters. This is a high stakes moment.

Afterlife

NOTES

p.7
Introduction

1. Paul Virilio, *The Original Accident*, first published in French as *L'Accident Originel* by Éditions Galilée in 2005 (Cambridge 2007).

p.22
Engineering at Home

1. Sara Hendren and Caitrin Lynch, *Engineering at Home* (2015), www.engineeringathome.org (accessed 5 August 2016).
2. Kathryn Allan and Djibril al-Ayad, 'Introduction', in Kathryn Allan and Djibril al-Ayad (eds), *Accessing the Future: A Disability-Themed Anthology of Speculative Fiction* (2015), pp.1–5, 3, http://press.futurefire.net/p/accessing-future.html (accessed August 2016).
3. Annemarie Mol, Ingunn Moser and Jeannette Pols (eds), *Care in Practice: On Tinkering in Clinics, Homes and Farms* (Basel 2010), p.14.
4. Hendren and Lynch 2015 (see note 1).
5. Allan and al-Ayad 2015, p.2 (see note 2).
6. Mol, Moser, and Pols (eds) 2010, p.14 (see note 3).

p.28
iPhone

1. *The Original iPhone Announcement Annotated: Steve Jobs' Genius meets Genius* (9 September 2015), https://thenextweb.com/apple/2015/09/09/genius-annotated-with-genius/ (accessed 29 August 2016).
2. Ibid.
3. *Apple Celebrates One Billion iPhones* (27 July 2016), http://www.apple.com/newsroom/2016/07/apple-celebrates-one-billion-iphones.html (accessed 29 August 2016).
4. Ian Bogost, *You are Already Living inside a Computer* (14 September 2017), https://www.theatlantic.com/technology/archive/2017/09/you-are-already-living-inside-a-computer/539193/ (accessed 3 October 2017).
5. Nir Eyal, *Hooked: How to Build Habit-Forming Products* (London 2014), p.1; and Sherry Turkle, *Reclaiming Conversation: the Power of Talk in the Digital Age* (New York 2015), p.42.
6. *Das Jugendwort des Jahres 2015*, http://www.jugendwort.de/ (accessed 5 July 2016).
7. Turkle 2015, pp. 21, 27 (see note 5).
8. 'The Adaptation of Work Rights to the Digital Era', Article 25, *Project de Loi visant à instituer de nouvelles libertés et de nouvelles protections pour les entreprises et les actifs* (24 March 2016), http://www.assemblee-nationale.fr/14/projets/pl3600.asp (accessed 29 August 2016).
9. Bruce Schneier, *Data and Goliath: the Hidden Battles to Collect your Data and Control your World* (New York 2015), p.34.
10. Sasha Issenberg, *How Obama's Team Used Big Data to Rally Voters* (19 December 2012), https://www.technologyreview.com/s/509026/how-obamas-team-used-big-data-to-rally-voters/ (accessed 11 September 2016).
11. James Williams, *The Clickbait Candidate* (3 October 2016), http://quillette.com/2016/10/03/the-clickbait-candidate/ (accessed 15 October 2017).
12. Benedict Evans, *16 Mobile Theses* (18 December 2016), http://ben-evans.com/benedictevans/2015/12/15/16-mobile-theses (accessed 5 July 2016).
13. Paul Lewis, *Our Minds have been Hijacked: the Tech Insiders Who Fear the Smartphone Dystopia* (6 October 2017), https://www.theguardian.com/technology/2017/oct/05/smartphone-addiction-silicon-valley-dystopia (accessed 9 October 2017).

p.34
Jibo

1. ELIZA was created at the MIT Artificial Intelligence Laboratory by Joseph Weizenbaum (1923–2008) between 1964 and 1966.
2. A chatbot, targeted at 18- to 24-year-olds in the USA, developed by Microsoft's technology and research and Bing teams to 'experiment with and conduct research on conversational understanding', see www.theguardian.com/technology/2016/mar/24/tay-microsofts-ai-chatbot-gets-a-crash-course-in-racism-from-twitter (accessed 7 July 2017). Having a Mac computer, there were not many computer games available, so I often played with ELIZA.
3. Front cover, *Discover* (October 1999).
4. See, for example, www.nytimes.com/2014/07/20/opinion/sunday/the-future-of-robot-caregivers.html?_r=0 (accessed 8 July 2017).
5. http://www.parorobots.com.
6. http://www.etymonline.com/word/robot.

p.56
Super Citizen Suit

1. http://www.wired.com/2016/01/social-media-made-the-arab-spring-but-couldnt-save-it/ (accessed August 2016); and https://www.theguardian.com/world/2011/feb/25/twitter-facebook-uprisings-arab-libya (accessed August 2016).

p.72
Aquila

1. Mark Zuckerberg on his Facebook page (26 September 2015), commenting on an appearance he made at the UN, https://www.facebook.com/zuck/posts/10102391777086341.
2. Guy Debord, *Theory of the Dérive* (Paris 1956).

p.78
BRCK

1. United Nations Human Rights Council, 'Report of the Special Rapporteur on the Promotion and Protection of the Right to Freedom of Opinion and Expression, Frank La Rue' (16 May 2011), http://www2.ohchr.org/english/bodies/hrcouncil/docs/17session/A.HRC.17.27_en.pdf (accessed 21 May 2016).
2. Personal communication.
3. International Telecommunication Union Telecommunication Development Bureau, 'ICT Facts & Figures: The World in 2015' (May 2015), http://www.itu.int/en/ITU-D/Statistics/Documents/facts/ICTFactsFigures2015.pdf (accessed 21 May 2016).
4. E. Hersman, 'Further Capital to Grow BRCK', BRCK blog (8 January 2016), http://www.brck.com/2016/01/further-capital-to-grow-brck/#.VODs0rST6zl (accessed 21 May 2016).

p.82
The Campaign to Stop Killer Robots

1. Human Rights Watch, Article 36, Association for Aid and Relief Japan, International Committee for Robot Arms Control, Mines Action Canada, Nobel Women's Initiative, PAX, Pugwash Conferences on Science & World Affairs, Seguridad Humana en América Latina y el Caribe (SEHLAC), and Women's International League for Peace and Freedom.

p.100
Tree Antenna

1. Margaret Atwood, 'Progressive Insanities of a Pioneer', *The Animals in That Country* (Boston 1968).
2. Ibid.
3. Suzanne Simard paper - https://www.nature.com/articles/41557
4. Ibid.
5. 'Performance of Trees as Radio Antenna', http:/dtic.mil/dtic/tr/fulltext/u2/742230.pdf, p.1.

6. Ibid.
7. Ibid., p.5.
8. Jalila Essaidi, *Bulletproof Skin, Exploring Boundaries by Piercing Barriers* (2012).

p. 104
ONKALO

1. 'Repository', Posiva, Finland, http://www.posiva.fi/en/final_disposal/final_disposal_facility/repository - . VzCzKWPqOpc> (accessed 9 May 2016).
2. Ibid.
3. Paul Virilio, *The Original Accident*, first published in French as *L'Accident Originel* by Éditions Galilée in 2005 (Cambridge 2007).
4. Ibid., p.9.
5. Paul Virilio, 'The Primal Accident', in Brian Massumi (ed.), *The Politics of Everyday Fear* (Minneapolis 2000), pp. 211–18, p.212.

p.110
Great Green Wall

1. See http://harpers.org/blog/2016/08/the-rainmakers-flood/ (accessed August 2016).
2. Wilber's quote is: 'Suppose (an army of frontier farmers) 50 miles, in width, from Manitoba to Texas, could acting in concert, turn over the prairie sod, and after deep plowing and receiving the rain and moisture, present a new surface of green growing crops instead of dry, hard baked earth covered with sparse buffalo grass. No one can question or doubt the inevitable effect of this cooling condensing surface upon the moisture in the atmosphere as it moves over by the Western winds. A reduction of temperature must at once occur, accompanied by the usual phenomena of showers. The chief agency in this transformation is agriculture. To be more concise. Rain follows the plow.' He also said: 'In this miracle of progress, the plow was the unerring prophet, the procuring cause, not by any magic or enchantment, not by incantations or offerings, but instead by the sweat of his face toiling with his hands, man can persuade the heavens to yield their treasures of dew and rain upon the land he has chosen for his dwelling... The raindrop never fails to fall and answer to the imploring power or prayer of labor.' For more, see Michael J. Harrower, 'Water Histories of Ancient Yemen in Global Comparative perspective', in M.J. Harrower, *Water Histories and Spatial Archaeology: Ancient Yemen and the American West* (New York, 2016), p.51; and Charles Dana Wilber, *The Great Valleys and Prairies of Nebraska and the Northwest* (Omaha 1881), p.70.
3. Later he served as Secretary of Agriculture under President

Grover Cleveland (1837–1908) in 1893, during which time Morton fired the rainmaker. On his department's payroll was General R.G. Dyrenforth (1844–1910), whose failed experiments led him to be called 'Dry-henceforth' in the press. See D.R. Hickey, S.A. Wunder and J.R. Wunder, 'J. Sterling Morton and Arbor Day', in D.R. Hickey, S.A. Wunder and J.R. Wunder, *Nebraska Moments*, 2nd edn (Lincoln, NE 2007), pp.121–8.

4. J. Sterling Morton, 'Arbor Day Address, April 22, 1887', in Nathaniel H. Egleston (ed.), *Arbor Day: Its History and Observance* (1896), pp. 22–7, http://www.foresthistory.org/blogs/Morton_address.pdf (accessed 30 April 2016).

5. Great Green Wall, https://en.wikipedia.org/wiki/Great_Green_Wall (accessed 2 May 2016).

6. Gavin Haines, '"Green wall" to Target Sahel Terrorism', *BBC News* (3 May 2013), http://www.bbc.co.uk/news/science-environment-22368945 (accessed 30 April 2106).

7. Ibid.

8. Mark Hertsgaard, 'A Great Green Wall for Africa?', *The Nation* (2 November 2011), http://www.thenation.com/article/great-green-wall-africa/ (accessed 30 April 2016).

9. Burkhard Bilger, 'The Great Oasis', *The New Yorker* (19 and 26 December 2011), http://www.newyorker.com/magazine/2011/12/19/the-great-oasis (accessed 1 May 2016).

10. Jacques Tassin, 'André Aubréville (1897–1982), a Pioneering Forester and a Visionary Mind', *Bois et forêts des tropiques* (2015), no. 323 (1), pp.7–19.

11. 'Timeline of Farming in the US, 1934–1940', *American Experience*, PBS, http://www.pbs.org/wgbh/amex/trouble/timeline/index_2.html (accessed 2 May 2016).

12. Tassin 2015, p. 9 (see note 10).

13. Ibid., p. 13.

14. Ibid.

15. Rosetta S. Elkin, 'Desertification and the Rise of Defense Ecology', *Portal 9*, issue no. 4 'Forest' (Autumn 2014), http://portal9journal.org/articles.aspx?id=130 (accessed 30 April 2016).

16. In Africa Aubréville had advocated planting eucalyptus in some places. The species is not native. It does grow fast, but requires much water and burns hot and fast. Its roots make it hard for other species to grow around it. In the 1970s during a drought in the Sahel, eucalyptus planting projects brought in industrial planting techniques, including bulldozers to clear land.

17. Hickey, Wunder and Wunder 2007, p. 128 (see note 3).

18. 'Afforestation in China: Great Green Wall', *The Economist* (23 August 2014), http://www.economist.com/news/international/21613334-vast-tree-planting-arid-regions-failing-halt-deserts-march-great-green-wall (accessed 30 April 2016).

19. Bilger 2011 (see note 9).

20. 'Afforestation in China: Great Green Wall', *The Economist* (23 August 2014) (see note 18).

21. Hickey, Wunder and Wunder 2007, p.121 (see note 3).

22. Byron Anderson, 'Biographical Portrait Julius Sterling

Morton', *Forest History Today* (Fall 2000), pp.31–3, http://www.foresthistory.org/publications/fht/fhtfall2000/morton.pdf (accessed August 2016).

23. Alpha Jallow, 'African Leaders Tackle Desertification to Counter Poverty, Insecurity' *Voice of America News* (4 May 2016), http://www.voanews.com/content/africa-desertification/3315228.html (accessed 9 May 2016).

24. 'Workshop on Monitoring and Assessment of Drylands: Forests, Rangelands, Trees and Agrosilvopastoral Systems', Food and Agriculture Organization of the UN (19 January 2015), http://www.fao.org/partnerships/great-green-wall/news-and-events/news-detail/en/c/274395/ (accessed 9 May 2016).

25. Nkechi Isaac, 'Biotechnology, Pivotal to Success of Great Green Wall Initiative – Minister', Leadership (19 April 2016), http://leadership.ng/news/519524/biotechnology-pivotal-success-great-green-wall-initiative-minister (accessed 9 May 2016).

26. 'Great Wall of Africa on the Cards', *Times Live* (5 May 2016), http://www.timeslive.co.za/thetimes/2016/05/05/Great-Wall-of-Africa-on-the-cards (accessed 9 May 2016).

27. Ibid.

p.120
Made In Space

1. Images of an ISS toolbox can be accessed via British astronaut Tim Peake's Flickr stream: https://www.flickr.com/photos/timpeake/albums/72157634315750824 (accessed 10 June 2016).

2. This was launched to the ISS on 21 September 2014 on the SpaceX CRS-4 resupply mission.

3. Mike Chen, 'How We "E-mail" Hardware to Space', *Backchannel* (17 December 2014), https://backchannel.com/how-we-email-hardware-to-space-7d46eed00c98#.osktdqro0 (accessed 10 June 2016).

4. Daniel O'Connor, 'Manufacturing in Space', *TCT Magazine* (17 March 2016), http://www.tctmagazine.com/3D-printing-news/manufacturing-in-space/ (accessed 12 June 2016).

5. Aamna Mohdin, 'Stephen Hawking: Humanity Will Only Survive by Colonizing Other Planets', *Quartz* (19 January 2016), http://qz.com/597326/stephen-hawking-humanity-will-only-survive-by-colonizing-other-planets/ (accessed 1 July 2016).

6. Mary-Ann Russon, 'Elon Musk: People Will Probably Die on the First SpaceX Missions to Mars', *IB Times* (14 June 2016), http://www.ibtimes.co.uk/elon-musk-people-will-die-first-spacex-missions-mars-1565387 (accessed 25 June 2016).

7. Ibid.

8. Personal conversation with David Delgado and Dan Goods, Visual Strategists at The Studio, at NASA's Jet Propulsion

Laboratory (11 August 2017).

9.   Ibid.

p.136
MnION

1.   Erika Hayden, 'Is the $1,000 Genome for Real?', *Nature* (15 January 2014), http://www.nature.com/doifinder/10.1038/nature.2014.14530 (accessed 29 April 2016).

2.   Erika Hayden, 'Pint-sized DNA Sequencer Impresses First Users', *Nature* (7 May 2015), vol. 521, pp.15–16, http://www.nature.com/doifinder/10.1038/521015a (accessed 1 April 2016).

3.   J. Parker et al., 'Field-based Species Identification of Closely-related Plants Using Real-time Nanopore Sequencing', *Scientific Reports* (2017), vol. 7, no. 1, Article: 8345, http://dx.doi.org/10.1038/s41598-017-08461-5. https://www.nature.com/articles/nbt.4060.

4.   Joshua Quick et al., 'Real-time, Portable Genome Sequencing for Ebola Surveillance', *Nature* (11 February 2016), vol. 530, pp.228–32.

5.   See 'MinION FAQS', *Oxford Nanopore Technologies* (2016), https://nanoporetech.com/community/minion-faqs (accessed 29 April 2016).

6.   The sequencing industry has far less self-regulation than the DNA synthesis (printing) industry.

7.   23andMe was co-founded in 2006 by Anne Wojcicki (ex-wife of Google co-founder Sergey Brin), Linda Avey and Paul Cusenza. As of March 2016, it reported more than 1.2 million customers, with over 80 per cent signed up for research participation. See '23andMe Enables Genetic Research for Research Kit Apps', *23andMe* (21 March 2016), http://mediacenter.23andme.com/blog/researchkit/ (accessed 27 April 2016).

8.   Charles Seife, '23andMe Is Terrifying, But Not for the Reasons the FDA Thinks', *Scientific American* (27 November 2013), http://www.scientificamerican.com/article/23andme-is-terrifying-but-not-for-the-reasons-the-fda-thinks/ (accessed 26 April 2016).

9.   D. Gershgorn, 'The FDA's 23andMe Decision Will Also Change the Rules For All At-home Medical Genetic Testing, *Quartz* (2017), https://qz.com/953486/fda-23andme-decision-at-home-medical-genetic-testing/.

10. Antonio Regalado, 'J. Craig Venter to Offer DA Data to Consumers', *MIT Technology Review* (22 September 2015), https://www.technologyreview.com/s/541516/j-craig-venter-to-sell-dna-data-to-consumers/ (accessed 25 April 2016).

11. Antonio Regalado, 'For $999, Veritas Genetics Will Put Your Genome on a Smartphone App', *MIT Technology Review* (4 March 2016), https://www.technologyreview.com/s/600950/for-999-veritas-genetics-will-put-your-genome-on-a-smartphone-app/ (accessed 19 April 2016).

12. Sequencing around 70,000 people, *The 100,000 Genomes Project* is owned by the UK's Department of Health and is the world's largest public sequencing project of its kind. See 'The 100,000 Genomes Project', *Genomics England* (2016), https://www.genomicsengland.co.uk/the-100000-genomes-project/ (accessed 28 April 2016).

13. Claire Maldarelli, '23andMe Discloses Police Requests for Customers' DNA', *Popular Science* (22 October 2015), http://www.popsci.com/23andme-publishers-transparency-report-that-reveals-authority-dna-requests (accessed 22 October 2015).

14. Carl Zimmer, 'Craig Venter's Health Nucleus Tries to Reshape Medicine', *Stat News* (5 November 2015), https://www.statnews.com/2015/11/05/geneticist-craig-venter-helped-sequence-the-human-genome-now-he-wants-yours/ (accessed 25 April 2016).

15. 'Veritas My Genome', *Veritas Genetics* (2016), https://www.veritasgenetics.com/mygenome-panel#sec-2 (accessed 28 April 2016).

16. The UK insurance industry has extended its moratorium until 2019. See Philippa Brice, 'UK Moratorium On Use of Genetic Tests by Insurers Extended', *PHG Foundation* (21 January 2015), http://www.phgfoundation.org/news/16536 (accessed 28 April 2016).

17. 'Metrichor: From Metric (Measure) and Ichor (Magical Blood)', *Metrichor: An Oxford Nanopore Company* (2016), https://metrichor.com/s/about.shtml (accessed 28 April 2016).

18. Liat Clark, 'Oxford Nanopore: We Want to Create the Internet of Living Things', *Wired UK* (24 April 2015), http://www.wired.co.uk/news/archive/2015-04/24/clive-brown-oxford-nanopore-technologies-wired-health-2015 (accessed 25 April 2016).

19. Erika Hayden, 'Cancer Blood Test Venture Faces Technical Hurdles', *Nature* (12 January 2016), http://www.nature.com/doifinder/10.1038/nature.2016.19152 (accessed 29 April 2016).

20. Clark 2015 (see note 18).

p.142
Cryonics

1.   B. Franklin and W.T. Franklin, *Memoirs of the Life and Writings of Benjamin Franklin*, Vol. III (Philadelphia 1809), pp. 402–3.

p.148
DeepMind

1.   Alan Turing, 'Computing Machinery and Intelligence', *Mind* (1950), vol. 49, pp.433–60, http://www.csee.umbc.edu/courses/471/papers/turing.pdf (accessed March 2016).

2. Volodymyr Mnih et al., 'Human-level Control through Deep Reinforcement Learning', *Nature* (2015), vol. 518, pp.529–33, http://www.csee.umbc.edu/courses/471/papers/turing.pdf (accessed March 2016).

3. David Rowan, 'DeepMind: Inside Google's super-brain', *Wired UK* (July 2015), http://www.wired.co.uk/article/deepmind (accessed March 2016).

4. Natalie Wolchover, 'Concerns of an Artificial Intelligence Pioneer', *Quanta Magazine* (21 April 2015), https://www.quantamagazine.org/20150421-concerns-of-an-artificial-intelligence-pioneer/ (accessed March 2016).

5. Raffi Khatchadourian, 'The Doomsday Invention', *The New Yorker* (23 November 2015), http://www.newyorker.com/magazine/2015/11/23/doomsday-invention-artificial-intelligence-nick-bostrom (accessed March 2016).

6. Nick Bostrom, *Superintelligence: Paths, Dangers, Strategies* (Oxford University Press, 2014), e-book location 6029.

7. Meeting with Stuart Russell, 2 December 2015.

8. Russell and Tegmark, 'An Open Letter', *Future of Life Institute* (2015), http://futureoflife.org/ai-open-letter/ (accessed March 2016).

9. Stuart Russell, Daniel Dewey and Max Tegmark, 'Research Priorities for Robust and Beneficial Artificial Intelligence', *Association for the Advancement of Artificial Intelligence* (Winter 2015), http://futureoflife.org/data/documents/research_priorities.pdf (accessed March 2016).

10. Ibid.

11. Ibid.

ACKNOWLEDGEMENTS

Work on *The Future Starts Here* started many years ago, and there are many people to thank for helping us get to 'here'.

Thank you to Volkswagen Group for their generous support of the exhibition, for their key loan of the autonomous car prototype, and for their enthusiasm for the project from an early stage.

We would like to thank all the lenders and contributors to the exhibition: 38 Degrees, 3Scan, ALE Co. Ltd., Alvar Aalto Museum, Julian Assange, Basque Coast UNESCO Global Geopark, Yves Béhar and Fuseproject, Bento Bioworks, Drew Berry and the Walter and Eliza Hall Institute, Better Shelter with IKEA Foundation and United Nations High Commissioner for Refugees, Sampriti Bhattacharyya and Hydroswarm, Bloomberg L.P., Blueroom, Bosch, Cynthia Breazeal and Jibo Inc., Brick by Brick and Croydon Council, Martin John Callanan, Cambridge Analytica, Campaign to Stop Killer Robots, Cryonics Institute, Heather Dewey-Hagborg, Mathias Disney at University College London, Jalila Essaïdi and BioArt Labs, Facebook, Fixperts, Foster + Partners, Google, Google DeepMind, the Greek Embassy in London, Happiest Baby, Sara Hendren and Caitrin Lynch, Alexi Hobbs, Indian Space Research Organisation, Interboro, Lucas Jones, Cesar Jung-Harada, Alex Kalman and Mmuseumm, Kickstarter, Kuehn Malvezzi, Lavazza, Little Sun, Long Now Foundation, LSE Library, Made In Space, Marius Ursache, MAXXI (National Museum of the XXI Century Arts), Julian Melchiorri, Antanas Mockus and Corpovisionarios, Greg More and OOM Creative, Mother Dirt, Nano Quantum Information Electronics, Nanotronics, Natural History Museum London, New European Media, the New Zealand government, NordGen, Norwegian Ministry of Local Government and Modernisation, Simon Nummy, Ocean Agency, Össur UK, Oxford Nanopore Technologies, Parabon, PARO Robots, Phonvert, QD Laser Inc. and the University of Tokyo Institute for the Qutaish family, Refugee Open Ware, Reversible Destiny Foundation, Rosa Labs, Anders Sandberg, Schiebel, Seismic, Hanif Shoaei, Taryn Simon and Gagosian Gallery, SpaceX, Speedo, Jonas Staal and The Kurdish Women Liberation Group, Studio Tomás Saraceno, Superflux, Superpedestrian, TeleGeography, Tesla, The Collective, The Living, The Seasteading Institute, UberEATS, UK Electoral Commission, University of Southampton, Pauline van Dongen with Holst Centre, Marina van Goor, Virgil Văduva, Volkswagen, what3words, Sam Woolley, YouGov, Zones Urbaines Sensibles.

A special thanks to Adrian Lahoud, Kamil Hilmi Dalkir and Michaela French at the Royal College of Art, Miranda July with Oumarou Idrissa, Kei Kreutler and the Libre Space Foundation, Rosetta Elkin, Ruth Ewan, Siddharth Srivastava and team at the University of California Berkeley, Skylar Tibbits and the MIT Self-Assembly Lab, Smout Allen, Stamen, Tellart, and Tom Tlalim, for creating new works for the exhibition.

Thank you to Nazes Afroz, Eric Striffler, Armin Linke Studio, Ian Webster, Final Cut for Real, Government Access Television for Cupertino, ITN/Hayley Barlow, Laura Poitras and the New York Times for permission to reproduce work in the exhibition.

Thank you to the exhibition advisors Arjun Appadurai, Alexandra Daisy Ginsberg, Sophie Hackford, Anders Sandberg and Susan Schuppli, who have brought a critical rigour to our curatorial premise.

Thank you to the contributing writers to this exhibition book – Arjun Appadurai, Teju Cole, Anne Galloway, Corinna Gardner, Alexandra Daisy Ginsberg, Jennifer Kabat, Alex Kalman, Natalie D. Kane, Justin McGuirk, An Xiao Mina, Richard Moyes, Anders Sandberg, Susan Schuppli and Leanne Shapton – each of whom have brought intelligence and sensitivity in making sense of these strange new things of the future. And to the many photographers and artists who have granted permission for the reproduction of their work here.

Thank you to our colleagues in the V&A department of Design, Architecture and Digital, who have encouraged and supported this exhibition from the start. We owe much to their public-spirited approach to design, and in particular to key objects originating from the Rapid Response Collecting programme. Kieran Long in his time as Keeper of the department was instrumental in the conception and curation of the exhibition and the editing of this volume. Thanks also to Christopher Turner, current Keeper of DAD, for his advice and support in the final stages.

Thank you to our colleagues in various V&A collection departments for their expertise and advice in selecting the nine historic objects that introduce each section of the exhibition: Johanna Agerman Ross, Rachel Church, Catherine Flood, Penelope Hines, Catherine Ince, Anna Jackson, Alexandra Jones, Tim Stanley and Salma Tuqan.

A huge thank you to our exhibition design team, who have developed a spatial concept that busts all the clichés of what the future should look like: architects Andrés Jaque, Laura Mora and Roberto Gonzalez at the Office for Political Innovation in Madrid, graphic designer Jeffrey Ludlow and team at 2x4 in Madrid, audio visual designers Anab Jain and Jon Ardern of Superflux, lighting designer David Robertson of DHA, and build contractors Hawthorn. In addition, the V&A's conservators and technicians have been central to this physical realization.

Thank you to Linda Lloyd-Jones and all those in the V&A Exhibitions department who have dedicated themselves to this project: Gemma Allen, Claire Everitt, Dina Ibrahim, Sarah Jameson, Samantha King, Phoebe Newman, Sarah Scott, and above all to Lauren Papworth, who has been unfailingly committed for the whole journey.

Thank you to Asha McLoughlin and Bryony Shepherd in the interpretation team for their intelligence and attention to detail in developing the exhibition narrative and gallery texts.

Thank you to Zara Arshad, research assistant to the exhibition and assistant editor of this volume. She has brought endless creativity, critical rigour and dedication to the project, and it could not have happened without her. And to Courtney Foote for her brief but invaluable contribution as intern.

Thank you to all those who have worked in producing this book. Coralie Hepburn and Tom Windross in V&A Publishing have always supported the ambitious ideas over the safe. Thank you to picture researchers Suzy Brogard, Fred Caws and Liz Edmunds for hunting down the many permissions; to Brendan Cormier for his comments on the draft; and to graphic designer Erwan Lhuissier of Julia for giving form to this volume and the voices it contains.

Thank you to the many other colleagues across the V&A who have contributed to the exhibition, book and associated programme, of which there are too numerous to mention. David Bickle, Sophie Brendel, Kati Price, Tim Reeve, Bill Sherman and Alex Stitt have all given vital support to the project at a senior level, ensuring it could be realized to the highest quality.

With sadness, we would like to thank the late Martin Roth, who as director of the V&A supported this project by committing it to the programme, made introductions to a number of lenders, and in particular offered a critical interrogation of our themes. His impact has made it undoubtedly stronger, and is evident in the final product. And finally, thank you to the present director Tristram Hunt, who has shown unfailing support for this project from day one of his tenure.

Rory Hyde and Mariana Pestana

CONTRIBUTORS

Anne Galloway

Anne Galloway is associate professor in the School of Design at the Victoria University of Wellington, New Zealand. Anne teaches undergraduate courses in Design Ethnography and Speculative Design, and leads the More-Than-Human Lab. In 2018–2019, she is conducting ethnographic research into human–sheep relations embodied through on-farm and in-laboratory care practices.

Corinna Gardner

Corinna Gardner is the V&A's senior curator of design and digital at the V&A, where she holds responsibility for contemporary product and digital design and leads the Museum's Rapid Response Collecting activities. In 2015, Corinna co-curated *All of This Belongs to You*, an exhibition about the design of public life, and prior to joining the V&A, she worked at the Barbican Art Gallery, on shows including *OMA: Progress*, *Bauhaus: Art as Life*, Random International's *Rain Room* and Cory Arcangel's *Beat the Champ*.

Mariana Pestana

Mariana Pestana is co-curator of *The Future Starts Here*. She works as a curator in the Design, Architecture and Digital Department at the V&A and is co-director of The Decorators collective. Mariana has previously lectured at Central Saint Martins, Chelsea College of Arts and Royal College of Arts. In 2013, she curated *The Real and Other Fictions* for the Lisbon Architecture Triennale and in 2016, *This Time Tomorrow* for the V&A at the World Economic Forum.

Leanne Shapton

Leanne Shapton is an author, artist and publisher based in New York City. She is the co-founder, with photographer Jason Fulford, of J&L Books, an internationally distributed not-for-profit imprint specializing in art and photography books. Shapton grew up in Mississauga, Ontario, Canada. Shapton's *Swimming Studies* won the 2012 National Book Critic's Circle Award for autobiography, and was long listed for the William Hill Sports Book of the Year 2012.

Justin McGuirk

Justin McGuirk is a writer and curator based in London. He is the chief curator at the Design Museum and teaches Design Curating & Writing at Design Academy Eindhoven. He has been the director of Strelka Press, the design critic of *The Guardian*, and the editor of *Icon* magazine. In 2012, he was awarded the Golden Lion at the Venice Biennale of Architecture for an exhibition he curated with Urban Think Tank. His book *Radical Cities: Across Latin America in Search of a New Architecture* is published by Verso.

Kieran Long

Kieran Long is the director of ArkDes, Sweden's national museum for architecture and design. He was previously Keeper of the department of Design, Architecture and Digital at the V&A, where he was part of the curatorial team for *The Future Starts Here*. He has written for many years on architecture, design and the public realm in newspapers and magazines and has presented two television series for the BBC.

Alex Kalman

Alex Kalman is a designer, writer, curator, filmmaker and creative director. He is the founder and director of Mmuseumm and the owner of What Studio? His projects have been exhibited in the Museum of Modern Art, the Metropolitan Museum of Art, the V&A and the Venice Architecture Biennale. His forthcoming book *Sara Berman's Closet* will be published by Harper Design in 2018.

Teju Cole

Teju Cole is a writer, art historian, and photographer. He is the Distinguished Writer in Residence at Bard College and photography critic of the New York Times Magazine. He was born in the US in 1975 to Nigerian parents, and raised in Nigeria. He currently lives in Brooklyn. He is the author of four books.

An Xiao Mina

An Xiao Mina is a technologist, writer and artist. She leads the product team at Meedan, where they build digital tools for journalists, and is an affiliate researcher at the Berkman Klein Center for Internet and Society at Harvard University. Mina is author of *Memes to Movements: How the World's Most Viral Media is Changing Social Protest and Power*.

Richard Moyes

Richard Moyes manages Article 36 – part of the Campaign to Stop Killer Robots and an Ethics and Society Partner of DeepMind. He previously established landmine clearance projects and has designed and developed international legal treaties providing protection from weapons.

Natalie D. Kane

Natalie D. Kane is curator of digital design at the V&A. Previously, she was curator of FutureEverything in Manchester, and co-curator of the 2017 edition of Impakt Festival, Utrecht. She is one half of curatorial research group Haunted Machines, and a visiting lecturer at the Institute of European Design.

Susan Schuppli

Susan Schuppli is an artist and researcher whose work examines material evidence from war and conflict to environmental disasters. She has published widely within the context of media and politics and is author of *Material Witness* (MIT Press, forthcoming). Schuppli is Director of the Centre for Research Architecture, Goldsmiths University of London and previously worked on the Forensic Architecture project.

Jennifer Kabat

Jennifer Kabat's essays have appeared in *Granta*, *Frieze*, *The Believer*, *Harper's* and *The White Review*, among others. She is working on a book exploring civic values from the modernist suburb where she grew up to where she lives now in the Catskill Mountains. She teaches contemporary art and theory and arts writing at New York University and The New School.

Zara Arshad

Zara Arshad is assistant curator of *The Future Starts Here*, a Trustee of the Design History Society, and founder of the Design China website. She has previously lectured at Royal College of Art, Central Academy of Fine Arts Beijing and Tsinghua University. From 2013 to 2015, she was the Friends of the V&A Scholar. Her research focuses on twentieth- and twenty-first-century design from East Asia, particularly China and Korea.

Alexandra Daisy Ginsberg

Alexandra Daisy Ginsberg is an artist, designer and writer. Her experimental practice explores values shaping design, science and technology. Daisy is the lead author of *Synthetic Aesthetics: Investigating Synthetic Biology's Designs on Nature* (MIT Press, 2014), and in 2017, finished 'Better', her PhD in Design Interactions at London's Royal College of Art, interrogating powerful promises of better futures.

Anders Sandberg

Anders Sandberg is senior research fellow at the Future of Humanity Institute, a part of the Oxford Martin School and the Faculty of Philosophy of University of Oxford. He has a background in computational neuroscience and is researching long-term futures, emerging technologies, reasoning under uncertainty and global catastrophic risks.

Rory Hyde

Rory Hyde is co-curator of *The Future Starts Here*, and the V&A curator of contemporary architecture and urbanism. He is adjunct senior research fellow at the University of Melbourne and a design advocate for the Mayor of London. His book *Future Practice: Conversations from the Edge of Architecture* is published by Routledge.

Arjun Appadurai

Arjun Appadurai is Goddard Professor of Media, Culture and Communication at New York University and is currently a Visiting Professor at Humboldt University in Berlin. He is internationally recognized for his work on globalization, cities and violence. He has also written extensively in planning, design and media. His most recent book is *Banking on Words: The Failure of Language in the Age of Derivative Finance* (University of Chicago Press, 2016). He is a member of the American Academy of Arts and Sciences.

ILLUSTRATION CREDITS